BIOLOGICAL SURVEY OF MITCHELL PLATEAU AND ADMIRALTY GULF, KIMBERLEY, WESTERN AUSTRALIA

Papers resulting from a joint field expedition to Mitchell Plateau, Kimberley, by the Western Australian Museum, Perth, and the Field Museum of Natural History, Chicago, U.S.A., 1976-1977

A Western Australian Museum Publication

PERTH 1981

Cover: *Livistona eastonii* (a fan palm) shown on shallow soil over massive laterite north-east of airfield, Mitchell Plateau. An open forest of *Eucalyptus miniata* and *E. tetrodonta* with *Livistona eastonii* understorey, the dominant species of the Plateau, is seen in the background. (Photo: K.F. Kenneally.)

ISBN 0–7244–8355–1

Published by the Western Australian Museum, Francis Street, Perth, Western Australia 6000.

Publication of this report would not have been possible without the financial assistance of AMAX Exploration (Australia), Inc., and the Trustees of the Western Australian Museum gratefully acknowledge that assistance.

CONTENTS

	Page
Part 1 General Introduction	
B.R. Wilson	
Introduction	1
Geology and Physiography	4
Climate	5
Early Biological Investigations	6
Narrative	8
References	11
Part 2 A Survey of the Vegetation and Flora of Mitchell Plateau, Kimberley, Western Australia	
R.J. Hnatiuk and K.F. Kenneally	
Abstract	13
Introduction	13
Vegetation Descriptions	16
Annotated List of Flora	52
Discussion	86
Acknowledgements	92
References	93
Part 3 A Survey of Riverside Mangrove Vegetation Fringing Tidal River Systems of Kimberley, Western Australia	
A.G. Wells	
Abstract	95
Introduction	95
Methods	97
Discussion	106
Acknowledgements	121
References	121
Part 4 Observations on Mammals of the Mitchell Plateau Area, Kimberley, Western Australia	
D. Kitchener, L.E. Keller, A. Chapman, N.L. McKenzie, A.N. Start, K.F. Kenneally	
Abstract	123
Introduction	123
Methods	125
Annotated Species List	127
Discussion	157
Acknowledgements	161
Appendix I – Vegetation	161
Appendix II – Trapline Sites	162
Appendix III – Habitats	166
Appendix IV – Young	167
References	168

Part 5 Birds of Mitchell Plateau and Adjacent Coast and Lowlands, Kimberley, Western Australia
R.E. Johnstone and L.A. Smith

Introduction	171
Annotated List of Species	184
Discussion	211
References	213

Part 6 Amphibians and Reptiles of Mitchell Plateau and Adjacent Coast and Lowlands, Kimberley, Western Australia
L.A. Smith and R.E. Johnstone

Introduction	215
Methods	216
Annotated List of Species	217
Discussion	225
Acknowledgements	226
References	226
Addenda	

Part 7 Freshwater Fish Fauna of the Mitchell Plateau Area, Kimberley, Western Australia
J.B. Hutchins

Introduction	229
Methods	229
Results	232
Annotated List of Species	234
Discussion	242
Acknowledgements	246
References	246

Part 8 Molluscan Fauna of the Admiralty Gulf, Cape Voltaire and the Institut Islands, Kimberley, Western Australia. Chitons, Meso- and Neogastropods
F.E. Wells

Introduction	249
Methods	249
Annotated List of Species	253
Discussion	261
Acknowledgements	262
References	262

Part 9 Zonation of Molluscs in a Mangrove Swamp in the Kimberley, Western Australia
F.E. Wells and S.M. Slack-Smith

Introduction	265
Methods	266
Results	266
Discussion	272
Acknowledgements	273
References	273

PART 1

BIOLOGICAL SURVEY OF MITCHELL PLATEAU AND ADMIRALTY GULF, KIMBERLEY, WESTERN AUSTRALIA: GENERAL INTRODUCTION

B.R. WILSON*

INTRODUCTION

The Mitchell Plateau lies in the Kimberley division of Western Australia on the divide between the Mitchell and Lawley Rivers, centred at approximately latitude 14°49′S, longitude 125°50′E. **(Figs 1, 2)**.

It is possibly the most remote and inaccessible part of Australia, and until the 1976 survey which is the subject of this report, little information on the flora and fauna of the region was available. Rich vegetation and abundant wildlife exist there still in near pristine condition and potable water is plentiful. Until only a few decades ago the area was inhabited by the Wunambal and Gambre Aboriginal people. Wunambal people were in the Mitchell River country until about 1954 and groups made lengthy visits to their traditional territories there well into the 1960s. (pers. comm. I. Crawford, Western Australian Museum.) Well preserved rock paintings exist in cliff-face caves and extensive kitchen middens are abundant around the coast, for example the southern side of Walsh Point at the mouth of Crystal Creek, testifying to the one-time presence of relatively large Aboriginal populations. Today, however, the Aboriginal people have left for the missions, cattle stations and towns to the east and south and the country remains virtually uninhabited. The only settlement in the vicinity is Kalumburu Mission at the head of Napier Broome Bay **(Fig. 1)** which may be reached by way of an unmade road from Gibb River Station.

A mining camp was established on the Mitchell Plateau in 1965 by Amax Bauxite (now Alumax Inc.) following the discovery there of potentially commercial deposits of bauxite, although so far only exploratory testing of the deposits has been attempted. The camp was serviced by an air-strip and a track connecting with the Gibb River-Kalumburu Road. A barge landing was established below the Plateau at Walsh Point in Port Warrender at the head of Admiralty Gulf. Other tracks constructed by the mining company made possible vehicle access to much of this previously inaccessible rugged country **(Fig. 2)**.

This newly created accessibility since 1965 has permitted biological investigations on a scale previously impractical. However when in the early 1970s the Western Australian Conservation through Reserves Committee proposed to undertake a

---

*Division of Natural Science, Western Australian Museum, Francis Street, Perth, Western Australia. 6000. Present address: National Museum of Victoria, Melbourne, Victoria. 3000.

review of the reserves system in the Kimberley (System 7) it was seen that although early reports indicated a biota of immense interest from the taxonomical, biogeographical and ecological points of view, basic information on the flora and fauna of the Mitchell Plateau region was lacking.

For this reason, in November 1976 a major biological survey was undertaken of Mitchell Plateau, the adjacent lands in the lower Mitchell and Lawley River drainage systems and the shores of Port Warrender. The survey was organised by the Western Australian Museum, with officers of the W.A. Department of Fisheries and Wildlife, the National Parks Authority and the Field Museum of Natural History (Chicago) also participating.

This survey report comprises papers resulting principally from the 1976 survey although they contain data from earlier and later fieldwork as well. The report also includes the results of a botanical survey conducted independently by staff of the Western Australian Herbarium and an account of the mangroves of Admiralty Gulf and elsewhere on the northern Kimberley coast by A.G. Wells of the University of Sydney. The latter study was made during a survey of saltwater crocodile populations aboard the R.V. *Harry Messel* in August 1977 (Messel & Burbidge, 1978).

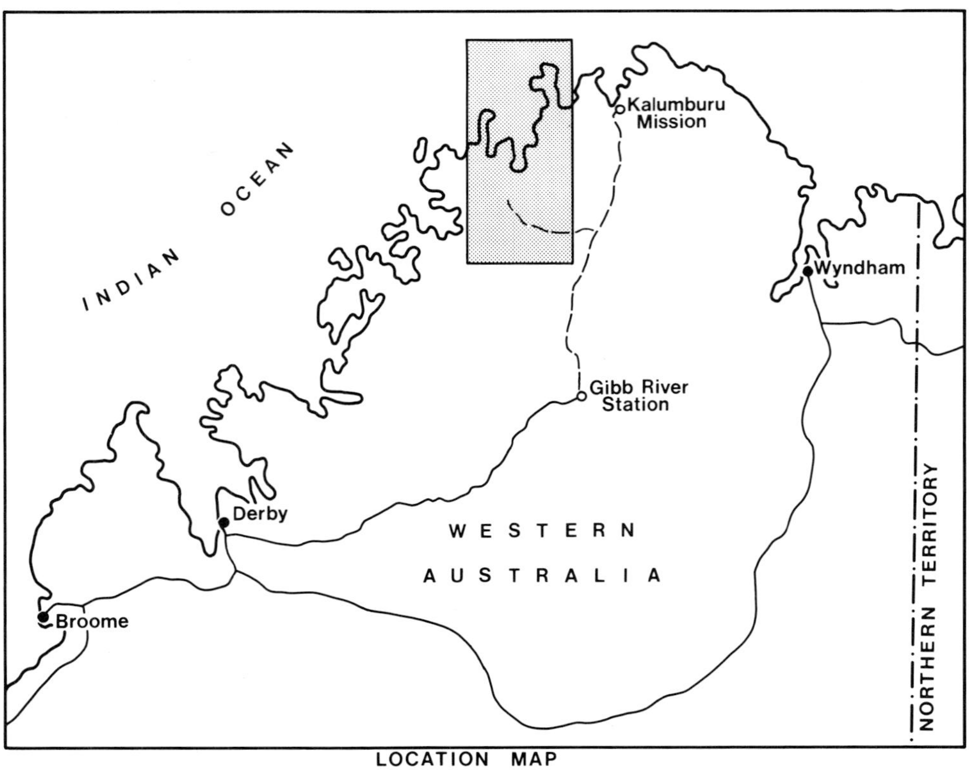

Fig. 1: Outline map of the Kimberley region.

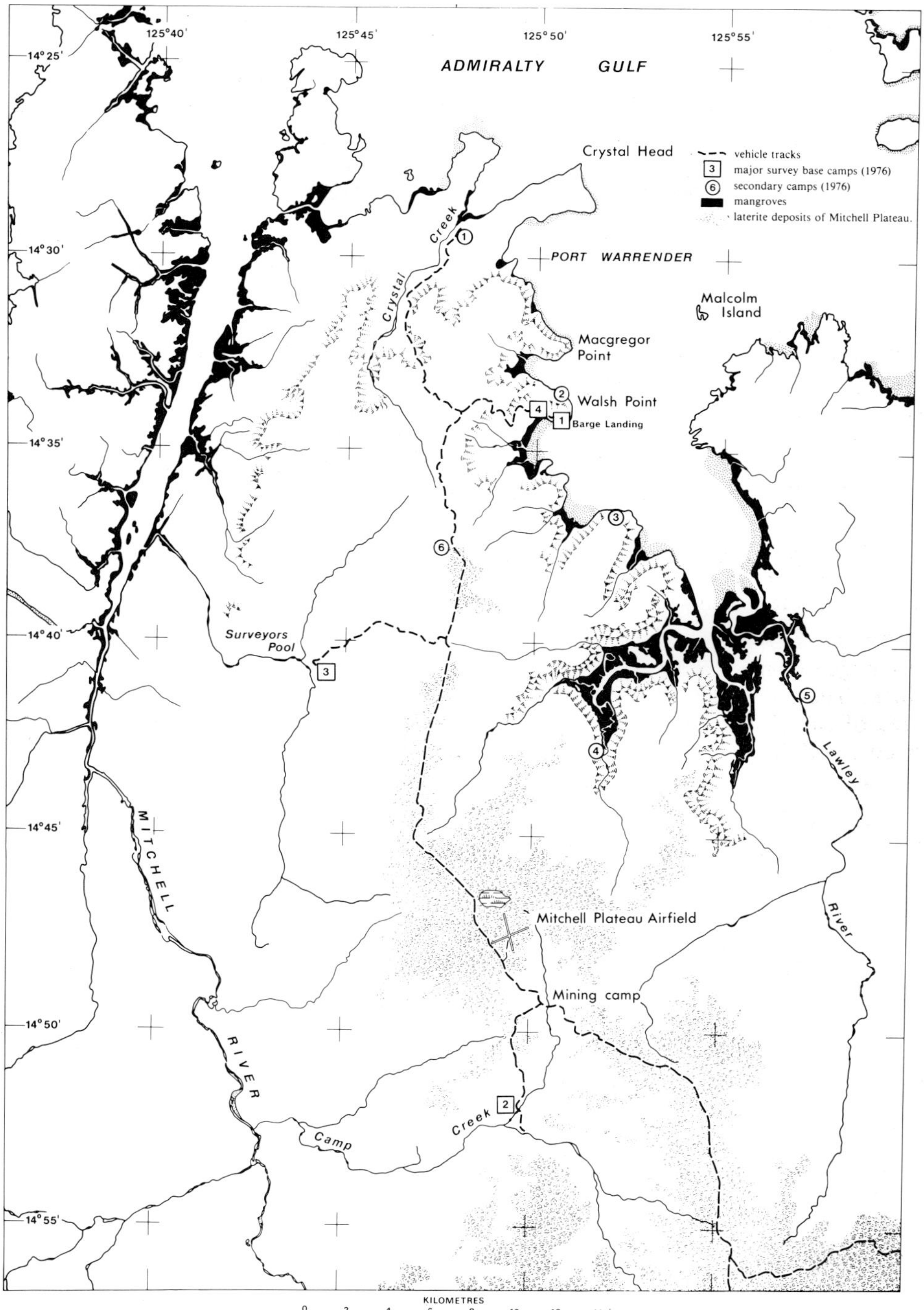

Fig. 2: The study area, Mitchell Plateau and adjacent lands and waters.

In addition to the taxa reported here, large collections were made of terrestrial mites, insects and other arthropods, terrestrial and freshwater molluscs, marine bivalve molluscs of Admiralty Gulf, and ectoparasites of vertebrates. Most of this material is currently being studied by relevant specialists and will provide material for publications in the *Records of the Western Australian Museum* and other journals. Most of the zoological specimens are deposited in the collections of the Western Australian Museum or the Field Museum of Natural History in Chicago. Botanical specimens are deposited at the Western Australian Herbarium.

GEOLOGY AND PHYSIOGRAPHY

The geology and physiography of the area encompassed by the Bureau of Mineral Resources Montague Sound 1:250,000 Sheet (SD 51-12) have been described by Allen (1966, 1971) and Sofoulis (1966).

Physiographically the area surveyed is part of the Kimberley Plateau Province. Two subprovinces are represented, the Prince Regent Subprovince and the Gibb Hills Subprovince. The Mitchell Plateau itself is a minor unit of the latter.

The Gibb Hills, formed on Carson Volcanics, occupy most of the area between the Mitchell and Lawley Rivers. They are characterized by rounded hills and a dendritic drainage pattern.

Above the 300 m contour along the divide between the Mitchell and Lawley Rivers drainage systems, there are remnants of superficial laterite deposits, as much as 15 m thick. These form the Mitchell Plateau. The surface of the plateau, which has an area of ca. 220 km^2 and an altitude of 300-350 m (Allen, 1971), is undulating and strewn with lateritic boulders. It is bordered by steep escarpments deeply dissected by stream erosion. Mesas and buttes are common.

West and east of the Gibb Hills are portions of the Prince Regent Plateau formed on strongly jointed, flat-lying beds of King Leopold Sandstone. On this surface a quite different topography has developed. Here, the extremely rough terrain is dominated by ridges of block-like sandstone boulders and narrow, steep-sided gorges formed along major joints, and the drainage system follows a rectilinear pattern controlled by jointing.

Both the Carson Volcanics and King Leopold Sandstones are members of the Kimberley Group, of Proterozoic (Carpentarian) age. The Carson Volcanics in the area surveyed are up to 300 m thick. They consist principally of massive and amygdaloidal basalt lavas. The King Leopold Sandstone is strongly bedded and jointed to the extent that it weathers into block-like outcrops with extremely deep horizontal and vertical fissures. Cross-bedding is also common. The quartz sandstone is fine to coarse-grained and is usually friable, but sometimes silicified and hard. Minor beds of fine conglomerate, pebbly sandstone and siltstone also occur.

The lateritic cap is best developed on the Carson Volcanics but shallow remnants also occur on the Sandstones. It is believed to be of Tertiary age.

There are several swamps on the plateau (**Fig. 2**). The largest of these is Airfield Swamp, about 80 ha in area and located approximately 5 km north of Mining

Camp. This fills with water in the wet season to a depth of at least 0.5 m and retains water at least until June (Beard, 1976). It was dry in September 1976.

The south-western side of Admiralty Gulf (Port Warrender) is characterized by steep-sided headlands (Gibb Hills) with rocky shores (Carson Volcanics) alternating with mangrove-lined bays fronted by wide intertidal mud-flats. The south-eastern side has a lower relief of craggy King Leopold Sandstone ridges and sandy beaches with fewer mangroves. At the southern head of the Gulf is the estuary of the Lawley River with a deep mangal drained by the Lawley itself and 2 major tidal creeks (Rail and Mindjau Creeks).

The western side of the study area is bordered by the Mitchell River and its long, narrow estuary. Much of the shore of this estuary is lined with mangroves or oyster-covered rocks and the hinterland is of low, craggy King Leopold Sandstone ridge.

CLIMATE

Slatyer in Speck *et al.* (1960) published information on the Kimberley weather patterns as known at that time. Beard (1976) updated this with data taken for the years 1968 to 1974 at the Mitchell Plateau Mining Camp and discussed the climate in context with the vegetation of the area. Additional weather data to 1978 have been obtained from the Bureau of Meteorology and Dr J. Gentilli of the University of Western Australia, and are summarised here in **Table 1** (data compiled by R. Hnatiuk and K. Kenneally).

In general terms, the climate may be described as 'dry tropical' (Bagnouls & Gaussen, 1957; Beard, 1976) or 'tropical savannah' (Koppen, 1936; Slatyer in Speck *et al.*, 1960), with distinct summer ("wet") and winter ("dry") seasons.

During the 10-year period of available records (**Table 1**) Mitchell Plateau had an average annual rainfall of 1583 mm. Heavy rainfall is usually experienced in the 4 months December to March and often in November and April as well. Heavy rain in October and May is uncommon and the winter (June to September) is usually rainless. Both the total annual rainfall and the length of the 'wet' season are variable from year to year. Using the Bagnouls and Gaussen formula, Beard (1976) calculated that there are usually six "ombrothermic" wet months each year in the Mitchell Plateau area; this ensures a longer growing season for plants than elsewhere in the Kimberley.

Temperature is consistently high at Mitchell Plateau with a registered annual mean maximum about 33°C. The highest monthly mean maxima (in excess of 35°C) occur from September to November but there is a distinct drop of several degrees once the summer rainy season begins. During the November 1976 biological survey, mid-day maxima of 44°C or more were experienced on several days. At the Mining Camp on the plateau some distance from the sea the temperature may fall to as low as 5°C during clear nights in the winter months.

During the 1970-1974 period, mean relative humidity was about 52% at 9 a.m. and 42% at 3 p.m. The range in monthly means (9 a.m. values) is from 74% in February to 31% in July.

Frost has not been recorded in the meteorological screen at Mitchell Plateau but low lying areas without tree cover could be expected to have ground-level radiation frosts during the dry season (pers. comm. Regional Director, Bureau of Meteorology, Perth).

Tropical cyclonic storms, which may be quite destructive to vegetation and shallow marine habitats, are common during the wet season at Mitchell Plateau. **Table 1** shows such storms in the immediate vicinity occur about twice in 3 years.

Table 1: A summary of some meteorological data from the Mitchell Plateau (data compiled by R.J. Hnatiuk & K.F. Kenneally).

		JUL	AUG	SEP	OCT	NOV	DEC	JAN	FEB	MAR	APR	MAY	JUN	YEAR
rainfall (mm)	$\bar{x}$	8	2	15	54	127	208	341	393	358	45	31	0.1	1 583
(Jan 1968-Dec 1977)	Sx	23	7	28	34	72	83	237	144	154	35	48	.3	341
ombrothermic "wet" months (Jan 1968-Dec 1977)		1	0	1	2	9	10	10	10	10	4	3	0	$\bar{x}$6
mean max. temp. (°C) (1970-1977)	$\bar{x}$	30.6	33.2	35.1	36.7	36.1	33.3	33.6	32.6	32.4	32.4	31.6	30.0	33.1
mean min. temp. (°C) (1970-1977)	$\bar{x}$	8.3	12.1	16.6	20.0	21.5	22.5	22.3	22.0	21.3	18.6	12.8	10.8	17.4
average temp. (°C) (= ½ (max. + mean)) (1970-1977)	$\bar{x}$	19.5	22.7	25.9	28.3	28.8	27.9	27.9	27.3	26.9	25.5	22.2	20.4	25.3
*no. cyclones/10 year period passing very near Mitchell Plateau (1920-1969)		-	-	-	-	-	7.0	1.9	1.9	1.6	0.5	-	-	0.66
*no. cyclones within 100 m of coast between Broome and Wyndham/10 year period (1920-1969)		-	-	-	-	0.2	1.8	4.2	3.4	3.0	0.4			1.3

*Personal communication, Dr J. Gentilli, Geography Department, University of Western Australia.

EARLY BIOLOGICAL INVESTIGATIONS

The French scientific expedition led by Nicholas Baudin on the *Naturaliste* passed the islands at the mouth of Admiralty Gulf in 1803 but apparently no biological collections or observations were made there by the expedition's naturalist François Peron.

Botanist Allan Cunningham participated in Phillip Parker King's surveys of the Kimberley coast aboard HMC *Mermaid* which anchored in Port Warrender on 11 and 12 October 1819. Cunningham accompanied a shore party which landed at Crystal Head to take bearings on Cape Bougainville. In King's narrative (1827) he refers to the sighting of a kangaroo there and several birds including 'tail-feathers of the *culus phasianus*' (Pheasant Coucal). On 11 and 12 October Cunningham accompanied a party in a whale-boat which explored Rail and Mindjau tidal creeks of the Lawley Estuary (King, 1827).

Although there is no reference in King's narrative to collection of biological specimens, Cunningham must have collected there or at Crystal Head because at least one plant species, the shrub *Emmenospermum cunninghamii* Bentham collected by Cunningham, has the type locality 'Port Warrender'. Cunningham's botanical collections are widely scattered, but a large number are at Oxford University. Others are at Kew, England, and in the N.S.W. National Herbarium and may include other material from Port Warrender. His journals for this period are in the Mitchell Library, Sydney, N.S.W.

Marine and terrestrial snails were also collected during King's surveys and some of these were deposited in the collections of the British Museum and subsequently described by Gray (1827), but there is no evidence that any of the specimens came from Admiralty Gulf.

The first zoological material to be reported accurately from this part of the Kimberley was collected by entomologist J.J. Walker while he was Chief Engineer aboard the survey ship H.M.S. *Penguin* in 1890 and 1891. Walker collected many insects, land-snails and other invertebrates which were deposited in the British Museum. The collections included material from Cassini, Condillac, Baudin and other islands near the mouth of Admiralty Gulf, and from Parry Harbour at the head of Cape Bougainville. The snails were subsequently described by Smith (1894).

In 1921 the 'Kimberley Expedition' led by William R. Easton travelled by horse and mule through the Mitchell Plateau area (Easton, 1922). The expedition members included Western Australian Government botanist C.A. Gardner. They travelled northwards along the eastern side of the Mitchell River (with one westward diversion across it to Swift Bay) to reach Crystal Head and rendezvous with a re-supply vessel. There the expedition divided, with one party travelling back south-west along the coast by sea and the other, including Gardner, continuing south and east, skirting the estuary of the Lawley River and eventually reaching the Drysdale River Mission (near the present site of Kalumburu). Gardner (1923) published a report of the vegetation along the route, including the Mitchell Plateau area, and his collections remain preserved in the Western Australian Herbarium.

In 1954 a survey party led by J.F. Morgan surveyed and constructed the roads to Kalumburu from Gibb River (Morgan, 1955). In the same year a party from the Division of Land Research and Regional Survey of the Commonwealth Scientific & Industrial Research Organization followed the same route. This party included N.H. Speck who reported on the vegetation (Speck, 1960) but although this publication is relevant, the route followed lies well to the east of Mitchell Plateau. In the same report, Speck published an account of northern Kimberley soils and geology, including land systems and geological maps derived from air-photo interpretation, while Stewart, Twidale and Bradley gave an account of the geomorphology, again based on air-photo interpretation.

Detailed geological investigations began at Mitchell Plateau in 1965 when a party from the Commonwealth Bureau of Mineral Resources and Geological Survey of W.A. worked there, using 4-wheel drive vehicle and helicopter. The results of this work were published as a 1:250,000 geological map sheet (Montague Sound, SD 51-12) with accompanying explanatory notes (Allen, 1966, 1971). Sofoulis (1966) reported on bauxite deposits.

Apparently, after Gardner's visit in 1922 there were no further biological investigations at Mitchell Plateau until the establishment of the Mining Camp in 1965. In June of that year, Professor A.J. Marshall and J.E. Nelson of Monash University visited the area and made small collections of animals including the Rock Rat, *Zyzomys woodwardi.* Amateur naturalist J.A. Smith was employed by the mining company from 1970 and he was responsible for sending the first zoological specimens and observations to the Western Australian Museum. Subsequently, with assistance from Smith and the mining company, Museum officers L.A. Smith and R.E. Johnstone (Jan.-Feb. 1973, June 1975) and Museum Honorary Associate W.H. Butler, (July-Aug. 1973, Oct. 1973, Sept.-Oct. 1978, Feb.-March 1979) worked in the area using 4-wheel drive vehicles, and made the first comprehensive zoological collections from the area.

Again with assistance from the mining company, botanist J.S. Beard spent a week at Mitchell Plateau in June 1974, examining the vegetation, especially patches of 'monsoon forest' which had been reported by company geologist K. Malcom (Beard, 1976).

Biological surveys conducted subsequently on the coastal islands (Burbidge & McKenzie, eds., 1978), and in the Prince Regent River Nature Reserve (Kabay & Burbidge, eds, 1977) have recently added a great deal of published information to knowledge of northern Kimberley biology but the flora and fauna of the mainland country south of Admiralty Gulf, including the Mitchell Plateau, was still little known when the joint museums' 1976 survey was conducted.

The area encompassed by this survey included Mitchell Plateau, the adjacent lands in the lower Mitchell and Lawley River drainage systems and the shores of Port Warrender.

NARRATIVE

1976 survey personnel:

Western Australian Museum team:

B.R. Wilson	leader
D.J. Kitchener, A, Chapman, E. Ellis A.N. Start (National Parks Authority), N.L. McKenzie (Dept. Fisheries & Wildlife) D.T. Kitchener (honorary)	mammology
F.E. Wells, S.M. Slack-Smith, C.W. Bryce	malacology
J.B. Hutchins, T. Duyker (honorary)	ichthyology
A.M. Douglas	entomology
L.A. Smith, R.E. Johnstone	ornithology & herpetology
A.F. Lovell	information officer
J.M. Chambal-Gaspar	photographer

Field Museum teams:

1. W.S. Street, J. Street, L.E. Keller, N.T. Allen (Univ. W.A.)	mammalogy
F.S. Lukoschus (Univ. Nijmegen, Netherlands)	parasitology
2. G.A. Solem	malacology
J.B. Kethley	entomology

Television camera crews from TVW Enterprises Ltd and Channel 9 Perth, and several newspaper journalists joined the field parties for brief periods.

The survey teams assembled at the Mitchell Plateau Mining Camp on 16 October 1976. The two Field Museum parties conducted a lengthy field programme in the Kimberley during the dry season of 1976, joining the Western Australian Museum team for the Mitchell Plateau survey in October-November. In their three 4-wheel drive vehicles they arrived by road via Gibb River. One Western Australian Museum 4-wheel drive vehicle also made the long journey by road from Perth. However, most of the Western Australian Museum personnel flew into the Mitchell Plateau airstrip by charter aircraft.

A 5-metre fibreglass boat equipped with an outboard engine was towed to Walsh Point by the Western Australian Museum vehicle. This was used for marine studies in Port Warrender and to transport personnel and equipment to coastal collecting stations on the southern side of the gulf.

Alumax Inc. generously supplied 2 additional 4-wheel drive vehicles from the Mining Camp fleet for use in the area. They also assisted in the provision of fuel and facilities.

After arrival in the early afternoon the 3 major field teams travelled by road to Camp Creek, Surveyors Pool and Walsh Point where base camps were established by nightfall. Trap-lines and out-camps were set up the following day.

On 28 October the base camps at Camp Creek and Surveyors Pool were broken. The Field Museum mammal group from Camp Creek shifted to a new site by the track at the base of Walsh Point peninsula. Western Australian Museum mammalogists who had also been at Camp Creek, and the team from Surveyors Pool, shifted to sites in the vicinity of Walsh Point.

During the final week of the survey, the fish team (Hutchins and Duyker) and the birds and reptiles team (Smith and Johnstone) operated independently of the other groups with vehicles visiting many localities in the Mitchell Plateau area.

All camps were broken on 5 November and personnel assembled at Camp Creek to pack and prepare for departure by vehicle and charter aircraft the following day.

Between 11 and 24 March 1977, three mammalogists revisited the area to obtain wet season data. This party consisted of A. Chapman and J. Henry from the Western Australian Museum and L.E. Keller from the Field Museum. They flew to Mitchell Plateau and were provided with a vehicle by Alumax Inc. They spent 14 days in the area collecting data from sites adjacent to the track on the Plateau, in the vicinity of Mining Camp, at the mouth of Crystal Creek, and at Walsh Point.

Locations

Australian Topographic Survey 1:100,000 maps were used for plotting locations. The Warrender (Series R611, 4068) and Admiralty Gulf (Series R611, 4069) sheets cover the area of the survey. Although many of the field location data were originally recorded as grid references based on these sheets all have been converted to latitude—longitude co-ordinates to comply with the Western Australian Museum's computer data-base standard. Positions of the base-camps and secondary camps are shown in **Fig. 2.**

1976 survey base-camps

1. South side of Walsh Point near the barge landing on the shores of Port Warrender (14°34′S, 125°50′E). Oct. 16-Nov. 5.
2. Camp Creek on the Plateau, approx. 5 km S.S.W. of the Mining Camp (14°52′S, 125°50′E). Oct. 16-24.
3. Surveyors Pool below the western escarpment of the Plateau (14°41′S, 125°44′E). Oct. 16-22.
4. Beside the track from the Mining Camp, 2 km from its terminus at Walsh Point (14°33′S, 125°49′E). Oct. 27-Nov. 5.

1976 secondary camps:

1. At the mouth of Crystal Creek (14°29′S, 125°48′E). D.J. & D.T. Kitchener. Nov. 1-4.
2. North side of Walsh Point (14°33′S, 125°50′E). D.J. & D.T. Kitchener. Oct. 27-30.
3. Point near Port Warrender (14°36′S, 125°51′E). A. Chapman, A. Douglas, N. McKenzie, A.N. Start. Oct. 28-Nov. 2.
4. Rail Creek, the western arm of the Lawley River estuary (14°41′S, 125°52′E). L.A. Smith & R.E. Johnstone. Oct. 18-27.
5. Head of the eastern arm of the Lawley River estuary i.e. mouth of Lawley River itself. (14°42′S, 125°57′E). J.B. Hutchins & T. Duyker. Oct. 20-27.
6. Beside the track about 25 km north of Mining Camp on the crest of the Plateau (14°37′S, 125°47′E). A. Solem & J.B. Kethley. Oct. 16-Nov. 5.

Fig. 2

Many local features in the study area were unnamed at the time of the 1976 survey. Consequently application was made to the Western Australian Nomenclature Advisory Committee to have a number of place names formally adopted to facilitate our work. These are:

Airfield Swamp	(14°46′S, 125°49′E)
Rail Creek	(14°41′S, 125°52′E)
Mindjau Creek	(14°42′S, 125°55′E)
Gurrbirri Creek	(14°40′S, 125°41′E)
Mertens Falls	(14°49′S, 125°42′E)
Glauert Creek	(14°56′S, 125°56′E)
Glauert Lagoon	(14°57′S, 125°57′E)

REFERENCES

ALLEN, A.D. (1966)—Geology of the Montague Sound 1:250,000 area (SD 51-12), Western Australia. *Rec. Bur. Miner. Resour. Aust.* no. 1966/20.

ALLEN, A.D. (1971)—*Explanatory notes on the Montague Sound geological sheet area.* Canberra: Bureau of Mineral Resources.

BAGNOULS, F. & GAUSSEN, H. (1957)—Les climats écologiques et leur classification. *Annls Geogr.* **66:** 193-220.

BEARD, J.S. (1976)—The monsoon forests of Admiralty Gulf, Western Australia. *Vegetatio* **31:** 177-192.

BURBIDGE, A.A. & McKENZIE, N.L. *eds* (1978)—The islands of the North-West Kimberley, Western Australia. *Wildl. Res. Bull. West. Aust.* no. 7: 1-47.

EASTON, W.R. (1922)—Report on North Kimberley of Western Australia. *Publs Dept. North-West., West. Aust.* no. 3.

GARDNER, C.A. (1923)—Botanical notes Kimberley Division of Western Australia. *Bull. For. Dept. West. Aust.* no. 22.

GRAY, J.E. (1827)—Mollusca; vol. 2, pp.474-496 in *Narrative of a survey of the intertropical and western coasts of Australia;* by P.P. King. London: John Murray.

KABAY, E.D. & BURBIDGE, A.A. *eds* (1977)—A biological survey of the Drysdale River National Park, North Kimberley, Western Australia in August, 1975. *Wildl. Res. Bull. West. Aust.* no. 6.

KING, P.P. (1827)—*Narrative of a survey of the intertropical and western coasts of Australia. Performed between . . . . 1818 and 1822.* 2 vols. London: John Murray.

KOPPEN, W. (1936)—Das geographische System der Klimate. *Handb. Klimat.* **1**(C).

MILES, J.M. & BURBIDGE, A.A. *eds* (1975)—A biological survey of the Prince Regent River Reserve, north-west Kimberley, Western Australia in August 1974. *Wildl. Res. Bull. West. Aust.* no. 3: 1-114.

MORGAN, J.F. (1955)—*Report on central North Kimberley region, North Kimberley survey and mapping expedition, 1954.* Perth: Dept. of Lands and Surveys, Western Australia.

SMITH, E.A. (1894)—On the land-shells of Western Australia. *Proc. malac. Soc. Lond.* **1:** 84-99, pl. 7.

SOFOULIS, J. (1966)—Bauxite deposits of the North Kimberley Region, Western Australia. *Ann. Rep. geol. Surv. West. Aust.* **1965:** p.69-70.

SPECK, N.H., BRADLEY, J., LAZARIDES, M., PATTERSON, R.A., SLATYER, R.O., STEWART, G.A. & TWIDALE, C.R. (1960)—The land and pastoral resources of the North Kimberley area, W.A. *Land Res. Ser. C.S.I.R.O. Aust.* no. 4: 1-109, pls. 1-7, 2 maps.

PART 2

A SURVEY OF THE VEGETATION AND FLORA OF MITCHELL PLATEAU, KIMBERLEY, WESTERN AUSTRALIA

R.J. HNATIUK*

and

K.F. KENNEALLY*

ABSTRACT

Mitchell Plateau, near the extreme north of Western Australia, is a lateritic plateau surrounded by basaltic slopes and rugged sandstone plains and gorges. Twenty-one alliances and sub-alliances are described and grouped under 14 structural formations—each with a list of plant species. Elements of special note include the understorey Fan Palm, *Livistona eastonii*, and remnant patches of vine thicket (rain forest) containing species of Gondwanan affinities. Documentation of the flora has resulted in 2 algae, 25 fungi, 29 lichens, 23 ferns and fern allies, 3 gymnosperms, 133 monocotyledons and 434 dicotyledons (total 649 species) being recorded.

INTRODUCTION

As part of the Western Australian Herbarium's study programme of the Kimberley region, three expeditions were made to document the vegetation and flora of Mitchell Plateau and surrounding areas. The results are the subject of this report, supplemented by information from valuable collections made at Mitchell Plateau by Beard (1976), Gardner (1923), George (1975, 1977), and by Western Australian Museum staff in 1976 and 1977.

Other botanical surveys of nearby parts of the Kimberley have been published by Miles and Burbidge (1975), Kabay and Burbidge (1977) and George (1978).

The first survey by us, in June 1976, documented the vegetation and flora of the plateau, sandstone and mangrove areas. The second, in May 1978, concentrated on the plateau, patches of remnant vine thicket (rain forest), and swamps. A third, in February 1979, was to collect flowering and fruiting material available only during the wet season.

Our visit in June 1976 followed a recent fire in which much of the understorey of the plateau vegetation had been destroyed. On the sandstone areas surrounding Surveyors Pool (**Fig. 1**), the fire had reached tree canopy level. By May 1978, however, a dense understorey dominated by grasses and herbs had grown to a height of about 1-1.5 m.

---

*Western Australian Herbarium, George Street, South Perth, Western Australia. 6151.

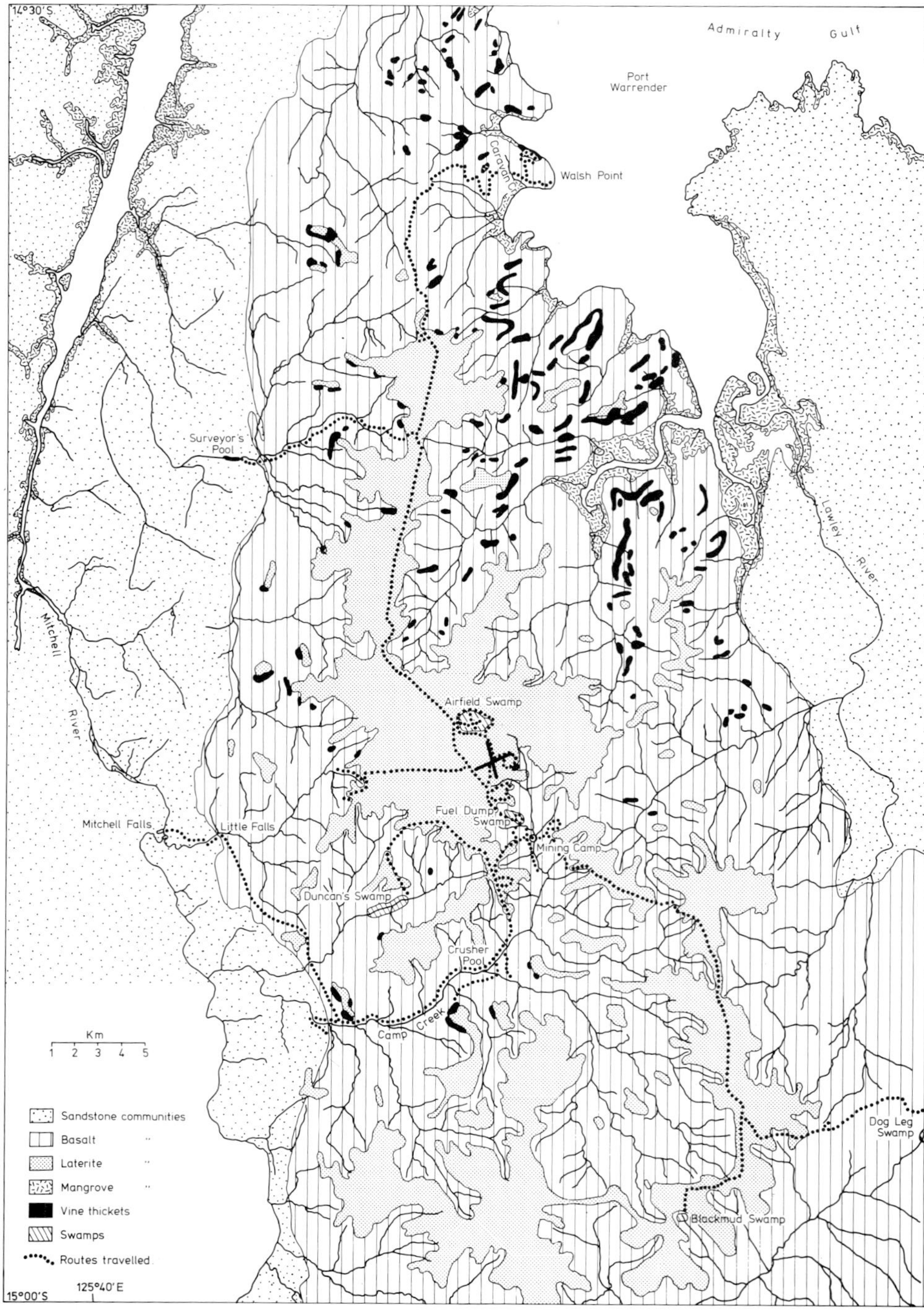

Fig. 1: Map showing localities, routes traversed, and general vegetation-geology of the Mitchell Plateau and surrounding area.

Table 1: Summary of vegetation units referred to in this paper. The numbers in brackets refer to the order in which they appear in the text.

Formation	Alliance	Association
Open-Forest	(1) *Eucalyptus tetrodonta-E. miniata*	
Low Open-Forest	(2) *Melaleuca leucadendron*	i. *Eleocharis brassii* ii. *Paspalum orbiculare* iii. *Eriocaulon quinquangulare/ E. setaceum*
Low Woodland	(3) *E. apodophylla*	
Woodland	(4) *E. tectifica-E. grandifolia: E. tectifica* sub-alliance	
Open-Woodland	(5) *E. tetrodonta-E. miniata: E. herbertiana* sub-alliance	
Low Closed-Forest	(6) Vine thickets (unclassified)	
	Mangroves:	
	(7) *Sonneratia alba-Avicennia marina* sub-alliance	
	(8) *Rhizophora stylosa* sub-alliance	
	(9) *Ceriops tagal-Avicennia marina* sub-alliance	
Low Woodland	(10) *Pandanus* spp.	
	(11) *Eucalyptus latifolia*	
	(12) *Eucalyptus confertiflora-E. polycarpa*	
	(13) *Terminalia fitzgeraldii*	
Low Open-Woodland	(14) *Terminalia canescens*	
	(15) *Eucalyptus papuana*	
Shrubland	(16) *Calytrix exstipulata-C. achaeta*	
Tall Grassland	(17) *Sorghum* sp. I	
Grassland	(18) *Sorghum* sp. II	
Low Grassland	(19) *Eriachne pauciflora*	
Closed-Herbland	(20) (unclassified)	
Open-Herbland	(21) *Portulaca clavigera-Calandrinia* spp.	
Miscellaneous unclassified communities	(22) Riverine (23) Cliff Faces (24) Aquatic Vegetation (25) Strand Line (26) Caravan Creek, Port Warrender.	

Our 1976 and 1978 visits followed soon after the summer wet season; on both occasions, Airfield Swamp contained water to a depth of about 1 m in the central part and many streams were still flowing.

The features of the climate most strongly affecting the vegetation are the distinct seasons: a relatively long 'wet' season (5-6 months) and a nearly rainless 'dry' season, the great variability in rainfall from year to year, the high frequency of tropical cyclones, and the great infrequency of frosts (which appear confined to the air very near the ground surface). Details of climate are given in Wilson (Part 1, this publication).

Climatic data (see **Table 1** p. 6) show that the centre of a cyclone could be expected to pass over or near the plateau twice in 3 years (or 1.5/yr.) and to occur within 160 km 13 times in 10 years (0.8/yr.). Heavy rains could be expected at the plateau from a cyclone passing within 160 km. With such a high frequency expected for cyclones, winds with sufficient force to be destructive can also be an expected part of the normal environment to which the vegetation has adapted. However, observations for this area do not yet permit an actual quantification of the frequency or extent of damage from severe storms at Mitchell Plateau.

Several geographic features of the Mitchell Plateau do not possess official names. In the absence of such names we have had to use our own which are enclosed in quotes in this report to indicate they are not officially accepted by the Nomenclature Advisory Committee of Western Australia.

VEGETATION DESCRIPTIONS

The vegetation of the Mitchell Plateau area is here divided into 21 alliances and sub-alliances, and 5 miscellaneous communities **(Table 1)**. The 21 units are grouped under 14 formations using the modified classification (Aplin, 1977) of Specht, Roe & Boughton (1974). Each unit is described below, and a list of species found in each is presented.

A generalized map of major vegetation complexes is presented in **Fig. 1**. The mapped units follow closely the major geologic boundaries with which they are easily identified in our area. Distribution data for *Eucalyptus* spp. comes largely from Hall *et al.* (1970-1977).

Open Forest Formation (trees 10-30 m tall, 30-70% cover)

1. *Eucalyptus tetrodonta-E. miniata* alliance **(Table 2, Fig. 2)**

The trees of this alliance are generally 15-18 m (range 12-19 m) in height. Projective foliage cover varies greatly depending upon soil depth and the availability of moisture, but is commonly in the range of 40-50%. Individual crowns of trees are generally either touching or separated by less than one crown width. The canopy trees comprise *Eucalyptus tetrodonta, E. miniata*, and *E. nesophila.* The proportions of these 3 species vary greatly from almost pure stands of *E. tetrodonta* to those with almost none of this species; e.g. *E. tetrodonta* is uncommon at the northern extremity of the plateau where *E. nesophila* is often the most abundant tree on the deep red clay overlying laterite. *E. miniata*, likewise, occurs with variable abundance; on the lateritic scree slopes of the plateau it appears more commonly than the other 2 species, although exceptions occur such as a pure stand of *E. miniata* on a small plateau remnant underlain by high grade gibbsite north-west of the main plateau.

Table 2: *Eucalyptus tetrodonta-E. miniata* alliance

TREES

Eucalyptus miniata
Eucalyptus nesophila
Eucalyptus tetrodonta

SMALL TREES

Erythrophleum chlorostachys
Gardenia sp. KFK 6746
Gardenia sp. KFK 6614
Grevillea pyramidalis
Livistona eastonii
Petalostigma pubescens

SHRUBS

Acacia ? *translucens* KFK 4889
Acacia sp. KFK 6659
Gossypium pilosum
Grevillea agrifolia
Grevillea heliosperma
Grevillea pyramidalis
Grewia polygama
Hibbertia sp. KFK 6739
Hibbertia sp. KFK 6807; Hnatiuk MP 47

HEMIPARASITES

Cassytha sp. KFK 6768

HERBS

Blumea integrifolia
Borreria sp. KFK 6605
Cartonema parviflorum
Cassia absus
Cassia mimosoides
Chlorophytum ? *laxum* ASG 14499
Commelina ensifolia
Cyclocarpa stellaris
Desmodium muelleri
Dipodium punctatum
Euphorbia sp. KFK 4911
Flemingia involucrata
Gomphrena sp. KFK 6667
Goodenia sepalosa
Gonocarpus leptothecus
Habenaria ferdinandii
Haemodorum sp. KFK 6656
Indigofera ? *pratensis* ASG 14470
Minuria sp. Hnatiuk MP 103
Mitrasacme connata
Mitrasacme elata
Pimelea sanguinea
Polycarpaea longiflora
Ptilotus conicus
Striga sp. KFK 4893, 6585
Tacca leontopetaloides
Tacca maculata
Tephrosia leptoclada
Tephrosia sp. KFK 6692, 5277
Vernonia cinerea
Zornia muriculata

GRAMINOIDS

Cenchrus elymoides
Coelorachis rottboellioides
Cymbopogon bombycinus
Eragrostis aff. *tenax* KFK 6594
Eriachne obtusa var. *major*
Heteropogon contortus
Themeda arguens
Themeda australis

FERNS

Cheilanthes ? *tenuifolia* KFK 6616

The understorey is dominated by *Livistona eastonii* palms that are mostly 8 m tall, but a great abundance of young palms characterizes many areas. Shrubs are very uncommon. The ground storey is dominated by grasses 1-2 m high. Many herbs and some dwarf shrubs also grow amongst the grasses. In the region of Mitchell Plateau, the alliance is most extensive on the lateritic plateau and its scree slopes. The *Livistona* understorey (socion or society) extends beyond the *E. tetrodonta-E. miniata* alliance on laterite and also occurs in parts of the *Eucalyptus confertiflora* alliance.

The *Eucalyptus tetrodonta-E. miniata* alliance is one of the most widespread in the Kimberley and extends eastwards across the Northern Territory to Cape York Peninsula in northern Queensland. It is found on a variety of soils.

Low Open Forest Formation (trees < 10 m tall, 30-70% cover)

2. *Melaleuca leucadendron* alliance **(Table 3)**

Tree height is variable, ranging from about 6-10 m. Tree cover is most commonly about 30-35% but some dense, small groves with more than 60% cover occur. *Melaleuca leucadendron* (*sens. lat.* Blake, 1968) is usually the sole species comprising the canopy.

The understorey of the *Melaleuca leucadendron* alliance varies from place to place. It may be described as separate associations characterized by their most prominent understorey species. Those distinguished in our area of study are:

a. *Eleocharis brassii* association, **(Table 3, Fig. 3)**. Found where the water is deepest and remains longest, e.g. the central area of Airfield Swamp.

b. *Paspalum orbiculare* association, **(Table 3, Fig. 4)**. Zone peripheral to the *Eleocharis* association but with slightly shallower water, e.g. Airfield Swamp. The socion characterizing this association is also found at 'Duncans Swamp' but without an overstorey of *Melaleuca leucadendron.*

c. *Eriocaulon quinquangulare-E. setaceum* association, **(Table 3)**. Shallow or deep water with dense *Melaleuca* overstorey, e.g. Airfield Swamp.

Table 3: *Melaleuca leucadendron* alliance

Table 3a: *Eleocharis brassii* association

TREES

Melaleuca leucadendron (sens. lat.)

HERBS

Crinum sp. KFK 4844a
Utricularia muelleri
Vallisneria aff. *caulescens* KFK 4761

GRAMINOIDS

Eleocharis brassii
Paspalum orbiculare

Table 3b: *Paspalum orbiculare* association

TREES

Melaleuca leucadendron

HERBS

Dysophylla verticillata

GRAMINOIDS

Ischaemum sp. KFK 7030
Panicum trachyrhachis
Paspalum orbiculare

Table 3c: *Eriocaulon quinquangulare-E. setaceum* association

TREES

Melaleuca leucadendron (sens. lat.)

HERBS

Dysophylla verticillata
Eriocaulon quinquangulare
Eriocaulon setaceum
Nymphoides ? hydrocharoides KFK 4826
Vallisneria aff. *caulescens* KFK 4761

The alliance is found where there is free standing water for many months of the year, including at least the first 3-4 months of the dry season (April-July). *Melaleuca leucadendron* is found across northern Australia and extends north into Malesia (*sensu* Steenis, 1951). It is usually associated with permanent (or nearly so) water, thus the distribution of the alliance may approach the species, but because there are still problems associated with the systematics of this species, a more precise statement of geographic range is not possible.

Low Woodland Formation (trees < 10 m tall, 10-30% cover)

3. *Eucalyptus apodophylla* alliance **(Table 4, Fig. 5)**

Trees reach a height of about 6-8 m. Canopy cover is about 10%, with some areas less than 10%, when the formation would be Low Open Woodland.

The understorey is herbaceous with a medium height grass layer (0.7-1.0 m), with 70% cover and a sparse to open low herb layer (1-30 cm).

The *E. apodophylla* alliance occurs on fine, light-grey, sandy silt that dries very hard during the rainless months. The soil is not freely draining and so is flooded for a portion of every year.

Table 4: *Eucalyptus apodophylla* alliance

TREES

Eucalyptus apodophylla
Eugenia eucalyptoides

SMALL TREES

Banksia dentata
Melaleuca viridiflora

SHRUBS

Grevillea pteridifolia

HERBS

Byblis liniflora
Buchnera linearis
Cassia mimosoides
Calochilus sp. KFK 7110
Cyclocarpa stellaris
Drosera indica
Goodenia pumilio
Indigofera sp. KFK 5314
Polygala leptalea
Stackhousia viminea
Utricularia chrysantha
Utricularia kimberleyensis
Utricularia limosa
Xyris complanata
Xyris aff. *indica* KFK 6721

GRAMINOIDS

Cyperus sp. KFK 5318 & 5319
Eriachne ciliata
Eriachne pauciflora
Fuirena ciliaris
Pseudopogonatherum contortum
Rhynchospora affinis
Sorghum aff. *brevicallosum* KFK 6636

The alliance is restricted to the extreme north-east of Western Australia and the north-west of the Northern Territory. It is especially rich in annual herbs that flourish in the range of moisture conditions that prevail at different times of the season.

Woodland Formation (trees 10-30 m tall, 10-30% cover)

4. *Eucalyptus tectifica* sub-alliance **(Table 5, Fig. 6)** of the *Eucalyptus tectifica-Eucalyptus grandifolia* alliance

The tree layer is composed most commonly of *E. tectifica* which grows to about 10-12 m (Woodland) on sites with alluvial soil, or to less than 10 m (Low Woodland) on sites with poor soil. Canopy cover varies from about 30% to<10% (Open Woodland) depending on soil conditions.

The Mitchell Plateau-Admiralty Gulf region is in the northern part of the *E. tectifica-E. grandifolia* alliance as described by Speck (1960). Here *E. grandifolia* is not a major component, being more typical of the southern part of the alliance. In our study area, *Eucalyptus foelscheana* is commonly associated with *E. tectifica* and may be recognized as a component association of the alliance. It occurs primarily on gentle slopes. Other trees which occasionally occur here are *E. confertiflora* on rugged topography, *E. latifolia* on shallow soil, and *E. nesophila* where laterite is shallow over basalt.

A sparse, low-tree layer is occasionally present and consists of *Hakea arborescens.* A dense ground layer of medium to tall grasses (1.3-2.0 m) occurs throughout the alliance. A few herbs occur amongst the grasses.

The alliance is largely confined to basaltic soils in our study area, which in general are of the 'fine textured yellow podsols' or 'Igneous Red Earths' of Speck (1960). Over its range, the alliance is found on a wide variety of soil types (sandstone, limestone and igneous derived soils).

The *E. tectifica* sub-alliance extends from Broome in Western Australia to Normanton near the south-east corner of the Gulf of Carpentaria, Queensland.

Table 5: *Eucalyptus tectifica* sub-alliance

TREES

Eucalyptus foelscheana
Eucalyptus tectifica

SMALL TREES

Acacia bidwillii
Acacia sp. KFK 6770
Cycas basaltica
Erythrina vespertilio
Hakea arborescens

SHRUBS

Acacia bidwillii
Acacia pallidifolia
Crotalaria linifolia
Gossypium pilosum

HERBS

Blumea mollis
Pygmaeopremna subacaulis

GRAMINOIDS

Heteropogon contortus
Sorghum sp. KFK 5205

Open Woodland Formation (trees 10-30 m tall, < 10% cover)

5. *Eucalyptus herbertiana* sub-alliance of the *E. tetrodonta-E. miniata* alliance **(Table 6, Fig. 7.)**

Tree height is about 10-12 m, canopy cover less than 10%. The canopy is very irregular due to wide spacing between trees as well as the variable, patchy nature of tree distribution. Another tree species associated with *E. herbertiana* is *Eucalyptus dichromophloia.*

Understorey shrubs are scattered in patches that are widely spaced due to patchy soil. The ground storey is also very irregular in distribution and density depending upon the availability of adequate pockets of soil.

Table 6: *Eucalyptus herbertiana* sub-alliance

TREES

Eucalyptus herbertiana
Eucalyptus dichromophloia
Eucalyptus latifolia

SMALL TREES

Acacia delibrata
Acacia gardneri
Brachychiton sp. KFK 7056
Cochlospermum fraseri
Ervatamia orientalis
Exocarpos latifolius
Ficus platypoda var. *platypoda*
Gardenia megasperma
Gardenia sp. KFK 5012
Grevillea pteridifolia
Hakea arborescens
Maytenus aff. *emarginatus* KFK 7129
Owenia vernicosa
Petalostigma pubescens
Planchonella pohlmanniana
Pouteria sericea
Terminalia carpentariae
Terminalia cunninghamii
Terminalia ferdinandiana
Terminalia hadleyana
Xanthostemon eucalyptoides
Xanthostemon paradoxus

SHRUBS

Acacia translucens
Boronia lanuginosa
Bossiaea bossiaeoides
Calytrix brachychaeta
Calytrix exstipulata
Crotalaria sp. KFK 5072
Distichostemon hispidulus
Exocarpos latifolius
Grevillea agrifolia
Petalostigma pubescens
Pouteria sericea
Rhynchosia sp. KFK 4807
Ricinocarpos cf. *marginatus* Maconochie 1277
Ricinocarpos sp. KFK 5007
Solanum beaugleholei
Solanum dioicum
Strychnos lucida
Triumfetta aff. *micrantha* KFK 7139
Triumfetta plumigera
Verticordia cunninghamii

CLIMBERS

Luffa graveolens

HEMIPARASITES

Decaisnina signata

HERBS

Crotalaria retusa
Crotalaria sp. KFK 5072
Desmodium ? *filiforme* KFK 4866
Gomphrena flaccida
Minuria sp. Hnatiuk/MP 103
Phyllanthus aff. *polycladus* W.V. Fitzg. KFK 5056
Polycarpaea longiflora
Ptilotus polystachyus f. *rubriflorus*
Tephrosia sp. KFK 5003
Tephrosia sp. J.Lewis 46
Trachymene dendritica
Trachymene didiscoides
Triumfetta plumigera

GRAMINOIDS

Plectrachne melvillei
Stipa sp. KFK 7044
Triodia sp. KFK 5047

The *E. herbertiana* sub-alliance is found on rugged sandstone country where soils are skeletal and confined to pockets and patches in the fissured and eroded rock surface. The soils are coarse, well-drained and low in nutrients.

The distribution of the sub-alliance is limited by the apparently disjunct range of *E. herbertiana*, in the extreme north-west of Western Australia and in Arnhem Land.

Low Closed Forest Formation (trees < 10 m tall, 70-100% cover)

6. Vine Thickets (**Table 7, Figs 8, 9, 10**)

Vine thickets are present in patches of from less than 1 ha to about 25 ha; since the canopy is closed about 5-10 m above ground (> 70% projective foliage cover) they can be classified as low closed forest (Specht *et al.*, 1974). Tall emergent crowns of some tree species (*Bombax, Sterculia, Terminalia, Brachychiton*) reach 10-15 + m above ground. Thus the profile of the vine thicket is characteristically very irregular in height and appears to consist of vine foliage as much as tree foliage. Most of the emergent trees are deciduous or semi-deciduous during the dry months from June to November. The understorey consists of low trees, shrubs (particularly in gaps below the tree canopy) and vines. There is virtually no ground flora apart from a few very young seedlings. Leaf litter is seldom more than 2-3 cm thick and frequently only gravel rubble and small mollusc shells are visible at the surface.

Table 7: Vine Thickets (* indicates deciduous species)

TREES

- * *Albizia lebbek*
- * *Bombax ceiba* var. *leiocarpum*
- * *Brachychiton paradoxum*
- *Canarium australianum*
- *Carallia brachiata*
- *Diospyros ferrea* var. *humilis*
- *Dysoxylum* sp. KFK 5134
- *Ganophyllum falcatum*
- *Garuga floribunda*
- *Lagerstroemia archerana*
- *Litsea* sp. KFK 5145
- *Myristica insipida*
- *Pouteria sericea*
- * *Sterculia quadrifida*
- * *Terminalia* sp. KFK 6706
- *Wrightia pubescens*

SMALL TREES

- *Acacia orthocarpa*
- *Aglaia* sp. KFK 5099, 5234
- *Albizia lebbek*
- *Atalaya ? hemiglauca* KFK 6708
- *Atalaya variifolia*
- *Brachychiton diversifolium*
- *Brachychiton paradoxum*
- *Canarium australianum*
- *Cassine melanocarpa*
- *Celtis philippensis*
- *Cochlospermum fraseri*
- *Denhamia obscura*
- *Diospyros nitens*
- *Dolichandrone heterophylla*
- *Dysoxylum* sp. KFK 5134, 6963
- *Euphorbia plumerioides*
- *Exocarpos latifolius*
- *Ficus opposita* var. *indecora*
- *Garuga floribunda*
- *Glochidion disparipes*
- *Lysiphyllum cunninghamii*
- *Melochia umbellata*
- *Mimusops elengi*
- *Planchonia australis*
- *Pouteria sericea*
- *Randia cochinchinensis*
- *Terminalia hadleyana*
- *Terminalia petiolaris*
- *Vitex glabrata*
- *Zizyphus quadrilocularis*
- *Zizyphus* sp. KFK 5265

Table 7 cont.

SHRUBS

? *Alectryon serratus* KFK 5103, 6675
Atalaya variifolia
Cadaba capparoides
Capparis sp. KFK 5107, 5231, 5343, 6677
Cassia glauca
Cassine melanocarpa
Celtis philippensis
Citriobatus spinescens
Clerodendrum floribundum
Clerodendrum aff. *ovalifolium* KFK 5152
Clerodendrum tomentosum
Crotalaria sp. KFK 6757
Dicliptera aff. *leonotis* KFK 5154, 6653
Exocarpos latifolius
Ficus opposita var. *indecora*
Ficus platypoda
? *Fissistigma* sp. KFK 7096
Glochidion disparipes
Glycosmis pentaphylla
Grewia sp. KFK 5128
Jasminum parviflorum
Melochia umbellata
Micromelum minutum
Murraya paniculata var. *ovatifoliolata*
Paramignya trimera
Phyllanthus sp. KFK 6711
Pouteria sericea
Pterocaulon verbascifolium
Trema sp. KFK 5112

CLIMBERS

Abrus precatorius
Canavalia papuana
Capparis quiniflora
Clematis pickeringii
Dioscorea transversa vel. aff. KFK 6660
Flagellaria indica
Jasminum parviflorum
Merremia sp. KFK 5120
Operculina sp. KFK 5123
Pachygone ovata vel. aff. KFK 5106; 5109; 5148; 5161
Passiflora foetida
Pisonia aculeata
Smilax australis
Thunbergia ? *arnhemica* ASG 14490
Trichosanthes sp. KFK 7010
Tylophora flexuosa

EPIPHYTES

Dendrobium dicuphum

HERBS

Achyranthes aspera
Asparagus racemosus
Hypoestes suaveolens
Pupalia lappacea

GRAMINOIDS

Cyperus sp. KFK 6756 (in small creek in v.t.)

The soils have been described as red or yellow earths developed on basaltic or lateritic scree (Beard, 1976).

Most vine thickets larger than 1 ha are found on the escarpment slopes of either the lateritic plateau or the massive, basaltic, lava flows overlooking Port Warrender. In these situations they may receive above-average amounts of ground water and in some cases may be protected from fires by the rock above or the boulder fields on which some grow. A few examples of vine thicket are also found in gorges in sandstone, e.g. near Mitchell Falls.

The vine thickets are relatively discrete vegetation and floristic units. Few of their species are found further away than the immediate fringing zone of tall closed grassland dominated by two species of *Sorghum: S. australiense* and *S.* sp. (undescribed).

Low Closed Forest Formation (trees <10 m tall, >70% cover)

7. *Sonneratia alba-Avicennia marina* sub-alliance of the Mangrove alliance **(Table 8, Fig. 11).**

Tree height is about 7-8 m and canopy cover about 80%, becoming less near the seaward margin. The canopy is deep (4-5 m), starting about 1.5 m above ground level, which presumably marks the high tide level. There are virtually no understorey plants except an occasional mangrove seedling. No intermediate size mangroves were seen here.

The *Sonneratia alba-Avicennia* sub-alliance marks the most seaward zone of mangroves in the study area. It measures about 125 m wide in the area south-west of Walsh Point. Aerial photos of the Admiralty Gulf indicate that the same sub-alliance occurs in this position around the Gulf and partway up tidal streams. The soil is a mixture of silt and fine organic matter bound together by a dense mat of mangrove roots. Pneumatophores (stout in *Sonneratia*, pencil-like in *Avicennia*) protrude in abundance from the soil.

Low Closed Forest Formation (trees <10 m tall, >70% cover)

8. *Rhizophora stylosa* sub-alliance of the Mangrove alliance **(Table 8, Fig. 11).**

Trees range from 7-10 m in height and canopy cover exceeds 80%. The canopy may take the form of a long, narrow ridge with tall trees forming the high, central point along the ridge and sloping down both landwards and seawards.

No understorey species of plants occur except for individuals or groups of recently established *Rhizophora* seedlings. The central trees are massive; landward ones are considerably smaller. No lichen, moss, or liverwort flora of any consequence, such as occurs in some *Rhizophora* forests in the world, grow on the trunks and branches.

The *Rhizophora* sub-alliance, in the Admiralty Gulf mangals, regularly forms a zone immediately landward of the *Sonneratia-Avicennia* sub-alliance.

Low Closed (open) Forest Formation (trees <10 m, >70% cover)

9. *Ceriops tagal-Avicennia marina* sub-alliance of the Mangrove alliance **(Table 8, Fig. 11).**

Trees reach only about 1.5-5 m above ground. *Ceriops* and *Avicennia* frequently appear segregated from each other into distinct, elongated patches, sometimes parallel to the shore line (the major alignment of mangrove zones apart from creek lines), but more often at right angles to the shore. From the aerial photographs it would appear that a complex of drainage and tidal inundation patterns probably controls this fine-scale pattern. In this sub-alliance in the study area it appears that *Avicennia* grows to about 4 m in height (as do a few *Rhizophora*) while *Ceriops* reaches only 1.5-2 m.

At the seaward side of this sub-alliance zone, an understorey of numerous seedlings of *Ceriops* and *Avicennia* grows, especially where the canopy is somewhat open. Associated with the seedlings is the understorey mangrove *Aegialitis annulata* which grows to 1 m.

Along the landward edge of the main mangrove zone, *Avicennia* reaches only 0.3-2 m. It then becomes associated with *Suaeda australis.* Occasional, small groves of *Avicennia* also occur at the extreme landward fringe of the mangal beyond the extensive mud flats. *Avicennia* and *Ceriops* extend around the lateral edges of the mangal in the bays of Admiralty Gulf. In these situations several other species of mangrove can be found, though seldom in any abundance: *Xylocarpus australasicus, Bruguiera exaristata* and *B. parviflora.*

Table 8: Mangroves

TREES

Rhizophora stylosa

SMALL TREES

Avicennia marina
Bruguiera exaristata
Bruguiera parviflora
Camptostemon schultzii
Ceriops tagal
Osbornia octodonta
Sonneratia alba
Xylocarpus australasicus

SHRUBS

Aegialitis annulata
Aegiceras corniculatum
Avicennia marina
Excoecaria agallocha
Suaeda australis

HEMIPARASITES

Amyema thalassium (on *Avicennia marina*)

Low Woodland Formation (trees <10 m tall, 10-30% cover)

10. *Pandanus* spp. alliance **(Table 9, Figs 12, 13).**

Trees grow to 6-7 m high but often are much lower. The canopy is very irregular in height and density, apparently resulting from a wide range of ages and modes of reproduction. At any one site there may be a few, isolated, tall individuals or a small grove of tall trees. Spread about their immediate vicinity are a host of smaller plants arising from either seed or stoloniferous shoots. Where several plants grow close together, there is frequently no understorey; more frequently there is an understorey of open or closed, medium-tall grasses, or more rarely, low herbs.

Table 9: *Pandanus* spp. alliance

TREES

Pandanus aquaticus
Pandanus darwinensis var. *latifructus*
Pandanus spiralis var. *convexus*
Pandanus spiralis var. *septem-loculatus*

SHRUBS

Burtonia subulata
Grevillea pteridifolia
Trema aspera

GRAMINOIDS

Coelorachis rottboellioides
Eriachne pauciflora
Pseudopogonatherum contortum
Sorghum plumosum

Two main habitats in our study area support this alliance. Bordering flowing creeks and permanent or nearly permanent pools are *Pandanus aquaticus* and *P. spiralis* var. *multimammillatus.* The soil, which is usually a coarse, alluvial sand, may be deep or may be confined to small pockets in bed rock such as along parts of fast flowing streams. The other major habitat, characterized by seasonal inundation and soils that are poorly-drained, light grey silts, supports groves of *Pandanus spiralis* var. *convexus, P. spiralis* var. *septem-loculatus*, and *P. darwinensis* var. *latifructus*. These species occur most commonly in shallow, gently sloping valleys near the headwaters of intermittent streams or, more rarely, in slight depressions on the lateritic plateau. In the latter habitat especially, the alliance is in close association with a surrounding *Eucalyptus*-dominated community, e.g. *E. tetrodonta-E. miniata* alliance. With further study, separate associations may be recognized for each of these habitat units.

Low Woodland Formation (trees <10 m tall, 10-30% cover)

11. *Eucalyptus latifolia* alliance **(Table 10, Fig. 14).**

Trees range from 6-9 m in height and grow in an open, park-like community. The understorey consists predominantly of low grasses (0.3 m) and small herbs.

In our study area, the *E. latifolia* alliance is largely confined to areas of shallow, lateritic podsols on the plateau. It is particularly abundant to the south and west of the airfield. Occasionally, *E. latifolia* trees are found on basaltic soils in valleys near the plateau, in association with *Eucalyptus tectifica, E, confertiflora,* and *E. foelscheana*. It also may be found intermixed with *E. miniata, E. tetrodonta* and *E. nesophila* around the margins of shallow soil areas on the plateau. Speck (1960) has recognized separate alliances for some of these combinations, but they are not distinct in our area.

E. latifolia occurs across the northern extremity of Western Australia and into north-western Northern Territory.

Table 10: *Eucalyptus latifolia* alliance

TREES

Eucalyptus latifolia

SMALL TREES

Buchanania obovata
Callitris intratropica
Cochlospermum fraseri
Planchonia australis
Terminalia canescens

SHRUBS

Acacia tumida
Acacia sp. KFK 6750
Acacia sp. KFK 6751

HERBS

Borreria aff. *geeingwa* ASG 14484
Calandrinia sp. KFK 6606
Calochilus holtzei
Gomphrena sp. KFK 6607
Lomandra tropica ssp. *tropica*
Polycarpaea corymbosa

GRAMINOIDS

Eriachne aff. *pauciflora* KFK s.n.

Low Woodland Formation (trees <10 m tall, 10-30% cover)

12. *Eucalyptus confertiflora-E. polycarpa* alliance **(Table 11, Fig. 15).**

Trees average about 5-6 m in height. Crown cover is variable but usually between 10 and 15%. Several low trees and shrubs grow in this alliance.

The ground storey consists of medium tall, medium to open growth of grasses and herbs which appears to vary in composition from place to place with changes in the nature of the soil, principally with regard to depth, texture and drainage.

The alliance, in our study area, occurs on the scree slopes of the lateritic plateau. It grades into the *Eucalyptus tectifica* sub-alliance where scree thins over basaltic soils, into the *Terminalia canescens* alliance where scree soils become thin over a basaltic base rock, and into the *Eucalyptus tetrodonta-E. miniata* alliance where laterite thins over basaltic soils. *Eucalyptus polycarpa* appears most commonly at moist sites within the alliance.

The *E. confertiflora* alliance is widespread across the north of Australia.

Table 11: *Eucalyptus confertiflora-E. polycarpa* alliance

TREES
- *Eucalyptus confertiflora*
- *Eucalyptus foelscheana*
- *Eucalyptus latifolia*
- *Eucalyptus polycarpa*

SMALL TREES
- *Erythrophleum chlorostachys*
- *Grevillea mimosoides*

SHRUBS
- *Acacia bidwillii*
- *Crotalaria* sp. KFK 6622
- *Calytrix achaeta*
- *Cochlospermum fraseri*
- *Ficus opposita* var. *indecora*
- *Gardenia* sp. KFK 5116
- *Grevillea refracta*
- *Grevillea mimosoides*
- *Indigofera* sp. KFK 5314
- *Petalostigma pubescens*
- *Terminalia canescens*
- *Waltheria indica*

HERBS
- *Haemodorum* sp. KFK 7062
- *Polycarpaea corymbosa*
- *Striga* sp. KFK 4776

GRAMINOIDS
- *Chrysopogon latifolius*
- *Heteropogon contortus*
- *Themeda australis*

Low Woodland Formation (trees <10m high, 10-30% cover)

13. *Terminalia fitzgeraldii* alliance **(Table 12, Fig. 16).**

Tree height averages 6-8 m and cover is variable within the 10-30% range for the Low Woodland formation. Other tree species found in the alliance are *Terminalia canescens, Eucalyptus bigalerita,* and *E. papuana.*

Table 12: *Terminalia fitzgeraldii* alliance

TREES
- *Eucalyptus bigalerita*
- *Eucalyptus papuana*

SMALL TREES
- *Acacia pallidifolia*
- *Terminalia canescens*
- *Terminalia fitzgeraldii*

HERBS
- *Euphorbia* sp. KFK 6775

GRAMINOIDS
- *Coelorachis rottboellioides*
- *Sorghum plumosum*
- *Sorghum* sp. KFK 6781

Soils are typically alluvial-colluvial, grey basaltics. They are moderately drained though in some locations may be inundated during the wet season. The alliance is not common in our study area and was found near 'Crusher Pool', on an alluvial plain east of Surveyors Pool, and in patches near the periphery of 'Duncans Swamp' **(Fig. 1)**.

The *Terminalia fitzgeraldii* alliance appears similar to Speck's (1960) *Terminalia* sp.-*Dichanthium foecundum* alliance. Further work may show them to be the same alliance.

Low Open Woodland Formation (trees <10 m tall, <10% cover)

14. *Terminalia canescens* alliance **(Table 13, Fig. 17).**

Trees are of various heights but commonly 3-6 m. Canopy cover is mostly less than 10% but may be more in some clumps of trees. The understorey contains few shrubs and disconnected patches of grasses and herbs which vary in extent depending upon the depth of soil.

The alliance is frequently encountered on shallow, skeletal, basaltic soils fringing open areas where base rock is exposed at the surface but is occasionally found on the laterite and sandstone. Because the soil in such localities is shallow and of very irregular depths and distribution, the vegetation is equally heterogeneous in stature and distribution. The composition of the vegetation is sufficiently consistent to permit its recognition as an alliance.

Table 13: *Terminalia canescens* alliance

SMALL TREES
- *Capparis umbonata*
- *Cochlospermum fraseri*
- *Terminalia canescens*

SHRUBS
- *Acacia suberosa*
- *Cochlospermum fraseri*
- *Gardenia* sp. KFK 6746
- *Grevillea heliosperma*
- *Petalostigma pubescens*
- *Terminalia canescens*

HERBS
- *Chrysogonum ecliptoides*
- *Ptilotus conicus*

GRAMINOIDS
- *Eriachne* sp. KFK 6797

Low Open Woodland Formation (tree height <10 m, >10% cover)

15. *Eucalyptus papuana* alliance **(Table 14, Fig. 18).**

Eucalyptus papuana trees are widely spaced, so that the community is really dominated by a tall grass stratum (2 m). However, the regular occurrence of the eucalypt is sufficient for us to use it as the characteristic species defining the alliance.

The alliance is most commonly found on alluvial soils but may occur at favourable sites below some vine thickets.

The alliance is widespread throughout the north Kimberley (Speck, 1960).

Table 14: *Eucalyptus papuana* alliance

TREES

Eucalyptus bigalerita
Eucalyptus papuana

SMALL TREES

Acacia bidwillii
Terminalia fitzgeraldii

HEMIPARASITES

Amyema bifurcatum

EPIPHYTES

Cymbidium canaliculatum

HERBS

Desmodium triquetrum
Euphorbia vachellii
Ptilotus conicus

GRAMINOIDS

Eragrostis sp. KFK 6594
Scleria sp. KFK 7024

Shrubland Formation (shrubs 1-2 m tall, 10-30% cover)

16. *Calytrix exstipulata-C. achaeta* alliance **(Table 15, Fig. 19).**

Shrubs grow 1-3 m tall. Where the majority of shrubs are >2 m tall the vegetation can be designated Tall Shrubland, otherwise it is Low Shrubland. The shrub canopy is uneven in height and density, largely as a result of variations in soil depth above solid rock. The understorey is patchy in distribution, again being related to uneven distribution of the shallow pockets and sheets of soil.

The alliance is best developed in clearings in the *Eucalyptus tetrodonta-E. miniata* alliance on the lateritic plateau. The massive laterite is either exposed at the surface or thinly covered with soil. A fragmentary representative of the alliance frequently occurs either at the very rim of the lateritic plateau where the soil is very scant and much rock is exposed, or in a similar situation but at the rim of gorges cut into the sandstone as found for example near Mitchell Falls and Surveyors Pool.

The alliance occurs across northern Australia.

Table 15: *Calytrix exstipulata-C. achaeta* alliance

SMALL TREES

Cochlospermum fraseri
Petalostigma pubescens
Planchonia careya

SHRUBS

Calytrix achaeta
Calytrix exstipulata
Grevillea heliosperma
Petalostigma pubescens

GRAMINOIDS

Plectrachne bynoei
Triodia pungens

Tall Closed-Grassland Formation (grass >2 m tall, >70% cover)

17. *Sorghum* sp. I alliance (**Table 16, Figs 8, 9, 20**).

A dense growth of *Sorghum* sp. reaches 2-3 m in height. The grass annually regrows from ground level. It either decomposes or is burnt each year. Only a few herbs and an occasional shrub occur.

The alliance may be a pyric-disclimax in many localities as there is circumstantial evidence indicating that if left unburned it would be replaced or overtopped by woody vegetation. This is particularly evident where it forms a belt 50-100 m wide around part or all of certain vine thickets (see Discussion section). The alliance occurs on red volcanic (basaltic) soils that are seasonally wet and moderately drained.

Table 16: *Sorghum* sp. I alliance

SMALL TREES

Albizia lebbek
Brachychiton paradoxum
Dolichandrone heterophylla
Sterculia quadrifida

SHRUBS

Albizia lebbek
Atalaya variifolia
Brachychiton paradoxum
Canarium australianum
Celtis philippensis
Clerodendrum floribundum
Dolichandrone heterophylla
Ficus opposita var. *indecora*
Lysiphyllum cunninghamii
Pouteria sericea

CLIMBERS

Abrus precatorius
Passiflora foetida

HERBS

Chrysogonum ecliptoides
Pterocaulon verbascifolium

GRAMINOIDS

Cenchrus elymoides
Sorghum australiense
Sorghum sp. KFK 5167

Closed-Grassland Formation (grass 1-2 m tall, >70% cover)

18. *Sorghum* sp. II alliance **(Table 17, Fig. 21).**

Dense stands of *Sorghum* sp. grow in areas of shallow soil that are submerged for 6-8 months of the year to a depth of usually less than 20 cm. An open understorey of aquatic herbs grows amongst the grasses. Occasionally a middle layer understorey composed of *Eriachne pauciflora* is found.

A good example of the alliance is found around the perimeter of Airfield Swamp. The soil here is 25-30 cm deep over massive laterite.

Table 17: *Sorghum* sp. II alliance

HERBS

Blumea integrifolia
Blumea sp. KFK 4750
Drosera burmannii
Drosera indica
Drosera petiolaris
Goodenia propinqua
Hemiarrhena plantaginea
Lobelia dioica
Mitrasacme laevis
Mitrasacme nudicaulis
Mitrasacme spp. KFK 6763, 4775
Nymphoides parviflora
Polygala leptalea
Stylidium spp. KFK 4754, 4762, 4821, 4843
Utricularia chrysantha
Utricularia lasiocaulis
Utricularia limosa
Xyris aff. *complanata* KFK 4860; 4851
Xyris aff. *elongata* KFK 4835
Xyris aff. *oligantha* KFK 6588
Xyris aff. *pauciflora* J. Lewis 25

GRAMINOIDS

Eriachne pauciflora
Fimbristylis sp. KFK 4836
Sorghum aff. *brevicallosum* KFK 6636

Low Closed-Grassland Formation (grass <1 m tall, >70% cover)

19. *Eriachne pauciflora* alliance **(Table 18, Fig. 22).**

Swards of *Eriachne pauciflora* develop on periodically flooded areas of shallow soil (10 cm deep) over solid rock. A sparse, dwarf-herb understorey may also develop.

This alliance was found as the outermost zone of Airfield Swamp and as a fringing zone between some areas of exposed basalt and the surrounding *Terminalia canescens* alliance.

Table 18: *Eriachne pauciflora* alliance

HERBS
- *Calandrinia* sp. KFK 6606
- *Cassia mimosoides*
- *Crotalaria humifusa*
- *Gomphrena canescens*

GRAMINOIDS
- *Eriachne pauciflora*
- *Sacciolepis myosuroides*

20. **Closed-Herbland Formations**

These need more detailed study. They occur in swamps and may form a variety of associations with alliances such as *Eucalyptus apodophylla* and *Pandanus* spp.

Open-Herbland Formations

21. *Portulaca clavigera-Calandrinia uniflora* alliance **(Table 19,** Figs 17, 23).

This is a rare alliance which occurs on very thin soils in crevices and shallow pockets in exposed basalt. Often, loose boulders about 10-30 cm diameter are prominent near the vegetated areas. Algae and lichens frequently cover the exposed rocks. As cyanophytic algae (free-living or in symbiosis as lichens) are common, they could be expected to provide a source of nitrogen to the herbland vegetation.

Table 19: *Portulaca clavigera-Calandrinia uniflora* alliance

HERBS
- *Calandrinia uniflora*
- *Cleome tetrandra*
- *Evolvulus alsinoides*
- *Heliotropium strigosum*
- *Heliotropium* sp. KFK 6788
- *Justago cleomoides*
- *Portulaca bicolor*
- *Portulaca clavigera*

GRAMINOIDS
- *Brachiaria holosericea*
- *Eriachne* sp. KFK 6797

Miscellaneous Communities

22. **Riverine Communities (Table 20, Fig. 24).**

The fringing vegetation of rivers and streams forms a distinctive feature of the area. However, it has not yet been studied sufficiently, and we do not wish to subdivide it into alliances at this stage. The Riverine and Vine Thicket communities have many species in common and may merge, as for example along portions of Camp Creek.

The floristic affinities of this assemblage are strongly Gondwanan and not Australian e.g. *Ficus racemosa* and *Ganophyllum falcatum.*

Table 20: Riverine communities

TREES
- *Alphitonia excelsa*
- *Calophyllum sil*
- *Canarium australianum*
- *Cryptocarya* sp. KFK 5028
- *Denhamia obscura*
- *Eugenia angophoroides*
- *Ficus coronulata*
- *Ficus racemosa* var *racemosa*
- *Ficus virens* var. *sublanceolata*
- *Ganophyllum falcatum*
- *Melaleuca leucadendron* (sens. lat.)
- *Metrosideros eucalyptoides*
- *Myristica insipida*
- *Terminalia platyphylla*
- *Timonius timon*
- *Tristania* sp. KFK 6764

SMALL TREES
- *Antidesma ghaesembilla*
- *Drypetes australasica*
- *Emmenospermum cunninghamii*
- *Eugenia eucalyptoides*
- *Melaleuca viridiflora*
- *Nauclea coadunatus*
- *Pandanus aquaticus*
- *Pandanus spiralis* var. *multimammillatus*
- *Piliostigma malabaricum*
- *Tristania* ? *suaveolens* ASG 13135
- *Tristania* sp. KFK 4952
- *Xanthostemon eucalyptoides*

SHRUBS
- *Acacia holosericea*
- *Acacia kelleri*
- *Acacia orthocarpa*
- *Dodonaea lanceolata*
- *Emmenospermum cunninghamii*
- *Ficus podocarpifolia*
- *Hibiscus* sp. KFK 5004
- *Jacksonia argentea*
- *Jussiaea suffruticosa*
- *Osbeckia australiana*

CLIMBERS
- *Passiflora foetida*
- *Smilax australis*

HERBS
- *Canscora diffusa*
- *Ludwigia parviflora*
- *Mitrasacme foliosa*

GRAMINOIDS
- *Arundinella nepalensis*
- *Coelorachis rottboellioides*
- *Dimeria ornithopoda*
- *Dimeria* sp. KFK 4923
- *Eriachne pauciflora*
- *Germainia truncatiglumis*

23. **Cliff Face Communities (Table 21, Fig. 25).**

Small soil pockets and overhanging cliffs provide habitats for such species as *Hoya australis.*

Insufficient is known about these communities and they cannot yet be classified.

Table 21: Cliff face communities

SMALL TREES
Pouteria sericea

SHRUBS
Calytrix exstipulata
Ficus platypoda var. *platypoda*
Hoya australis
Malaisia scandens
Pouteria sericea

CLIMBERS
Flagellaria indica

HERBS
Canscora diffusa
Psoralea leucantha

GRAMINOIDS
Fimbristylis arthrostyloides

FERNS AND FERN ALLIES
Adiantum philippense
Ceratopteris thalictroides
Dicranopteris linearis
Drynaria quercifolia
Lindsaea ensifolia
Nephrolepis hirsutula
Selaginella sp. KFK 5083
Stenochlaena palustris

24. **Aquatic Communities (Table 22).**

Permanent to semi-permanent pools often support species of aquatic plants. Pools have been found on a variety of substrates including basaltic soils, and sandstone. Again, insufficient information prevents classification.

Table 22: Aquatic communities

HERBS
Crinum asiaticum
Dysophylla verticillata
Eriocaulon quinquangulare
Eriocaulon setaceum
Eriocaulon sp. KFK 4827
Limnophila sp. KFK 4941; 5333; 6782a
Myriophyllum trachycarpum
Nymphaea gigantea
Nymphoides aff. *hydrocharoides* KFK 4758
Nymphoides aff. *parvifolia* KFK 4747; 4760
Najas graminea
Ondinea purpurea
Triglochin procera var. *dubia*
Tristicha trifaria
Utricularia exoleta
Utricularia muelleri
Vallisneria aff. *caulescens* KFK 4830

GRAMINOIDS
Panicum trachyrhachis

STONEWORTS
Chara leptopitys
Chara ? *vulgaris* Hnatiuk/MP 76

25. **Strand Line Communities (Table 23, Fig. 26).**

Limited observations revealed a number of common strand species but the strand community did not appear well developed. Further studies are needed.

Table 23: Strand line community

SMALL TREES
Euphorbia plumerioides
Thespesia populneoides

CLIMBERS
Canavalia rosea

SHRUBS
Avicennia marina
Suaeda australis

GRAMINOIDS
Cyperus spp. KFK 5215; 6642
Fimbristylis cymosa
Sporobolus virginicus

26. **Ephemeral Fresh Water Creeks (Table 24, Fig. 27).**

A fresh-water, seasonal stream 'Caravan Creek' flowing into Port Warrender, west of the landing site near Walsh Point, supports a distinctive suite of species whose status as a vegetation type is uncertain at present. The vegetation is characterized by *Adansonia gregorii* in the tree stratum and a dense growth of Cyperaceae species and *Typha* in the ground stratum.

Table 24: 'Caravan Creek', Port Warrender

SMALL TREES

Adansonia gregorii
Pandanus spiralis
Terminalia petiolaris

SHRUBS

Ludwigia parviflora
Psoralea badocana

CLIMBER

Diplocyclos palmatus

HEMI-PARASITE

Cassytha sp. Hnatiuk/MP 108

HERBS

Ammannia multiflora
Euphorbia sp. KFK 5194

GRAMINOIDS

Cyperus conicus
Eragrostis japonica
Fimbristylis cephalophora
Fuirena ciliaris
Fuirena umbellata
Lipocarpha microcephala
Typha domingensis

Table 25: Zonation of mangrove sub-alliances at Port Warrender. The zones are listed from the most seaward to the most landward.

Species	Average Width of Zone (metres)
1. *Sonneratia alba-Avicennia marina*	50
2. *Sonneratia alba*	75
3. *Rhizophora stylosa*	100
4. *Ceriops tagal-Rhizophora stylosa-Aegialitis annulata*	175
5. *Ceriops tagal-Avicennia marina*	50-150
6. Mud flats	Variable (1300-2000)
7. *Avicennia marina*	±20

Fig. 2: *Eucalyptus tetrodonta-E. miniata* alliance: *E. tetrodonta* is large tree on left, *E. nesophila* is large tree on right, and *E. miniata* is white barked tree in centre background. *Livistona eastonii* forms a dense understorey.

Fig. 3: *Melaleuca leucadendron* alliance: *Eleocharis brassii* association in the central area of Airfield Swamp.

Fig. 4: *Melaleuca leucadendron* alliance: *Paspalum orbiculare* association around the periphery of the *Eleocharis brassii* association at Airfield Swamp.

Fig. 5: *Eucalyptus apodophylla* alliance with *Melaleuca leucadendron* and *Pandanus* sp. in wettest areas.

Fig. 6: *Eucalyptus tectifica* sub-alliance on basaltic soils peripheral to the plateau. Some *E. foelscheana* occurs here as well.

Fig. 7: *Eucalyptus herbertiana* alliance on rough sandstone east of Mitchell Falls. *Terminalia hadleyana* is large tree on left with *Triodia* sp. in foreground.

Fig. 8: Vine thicket south of 'Crusher Pool' with emergent *Albizia lebbek*. A dense peripheral zone of the *Sorghum* sp. I alliance grows in foreground and a low thicket canopy with many lianes.

Fig. 9: Vine thicket south of 'Crusher Pool' five months after severe October (1978) fire that penetrated the thicket. *Sorghum* sp. I alliance grows in foreground.

Fig. 10: Vine thicket on NW side of Plateau. *Sterculia quadrifida* is the emergent, deciduous tree and *Garuga floribunda* is the large evergreen tree in the middle ground.

Fig. 11: Mangrove alliance south of Walsh Point, Admiralty Gulf. For zonation, see text. The seaward edge is on the left.

Fig. 12: *Pandanus* spp. alliance. *Pandanus aquaticus* forming a dense fringing stand along Camp Creek.

Fig. 13: *Pandanus* spp. alliance. *Pandanus spiralis* in closed grassland of 'Duncans Swamp'

Fig. 14: *Eucalyptus latifolia* alliance on shallow soil on Mitchell Plateau.

Fig. 15: ***Eucalyptus confertiflora-E. polycarpa*** alliance. East side of Camp Creek with *E. confertiflora* dominant, and *E. tectifica* occasionally occurring.

Fig. 16: *Terminalia fitzgeraldii* alliance on edge of closed grassland by 'Duncans Swamp'.

Fig. 17: *Terminalia canescens* alliance on shallow basaltic soils north of plateau. Exposed basalt and the *Portulaca-Calandrinia* alliance is seen in the foreground.

Fig. 18: *Eucalyptus papuana* alliance as it is found in closed-grassland north of 'Crusher Pool'. *Terminalia fitzgeraldii* grows as an open understorey.

Fig. 19: *Calytrix exstipulata-C. achaeta* alliance on shallow soil on laterite of Plateau.

Fig. 20: *Sorghum* sp. I alliance growing on red basaltic soil north of the plateau. Tallest grasses grow on deep, grader-produced mounds at edge of road. *Cycas basaltica* is seen left of centre.

Fig. 21: *Sorghum* sp. II alliance growing on shallow, winter wet soil around Airfield Swamp.

Fig. 22: *Eriachne pauciflora* alliance on shallow soil over massive laterite. Taller grasses are *Sorghum* sp. growing in pockets of deeper soil.

Fig. 23: *Portulaca clavigera-Calandrinia* spp. alliance on very shallow soil on exposed basalt east of Camp Creek.

Fig. 24: Riverine community along Camp Creek showing fringe of *Pandanus aquaticus* and a large tree of *Nauclea coadunatus.*

Fig. 25: Cliff face community on sandstone near Mitchell River Falls.

Fig. 26: Strand line vegetation at Port Warrender with *Canavalia rosea* and *Cyperus* spp. in foreground. A patch of vine thicket (with the strand tree species *Euphorbia plumeroides* and *Thespesia populneoides*) grows onto the beach.

Fig. 27: An ephemeral fresh-water creek near Walsh Point, Port Warrender, showing a dense stand of *Typha domingensis*.

Fig. 28: *Adansonia gregorii* on basalt above Port Warrender. Aboriginal stone arrangements in foreground.

ANNOTATED LIST OF FLORA

The following annotated list contains all the taxa so far identified. It totals 641 taxa, in which there are 2 stoneworts, 25 fungi, 29 lichens, 23 ferns and fern allies, 3 gymnosperms, and 559 flowering plants (133 monocotyledons, 434 dicotyledons). The monocotyledons belong to 74 genera in 23 families, and the dicotyledons to 225 genera in 82 families. Voucher specimens for each taxon are housed in the Western Australian Herbarium (PERTH).

The species list is arranged as follows. Fungi are arranged alphabetically within each class. Lichens are arranged alphabetically under genera. Ferns, gymnosperms, monocotyledons, and dicotyledons are arranged alphabetically within families which are themselves arranged alphabetically in each section. New records are indicated by a dagger (†). Uncertain determinations are followed by collectors' numbers, for easy reference if a supplementary flora list is published later.

An annotated flora of the Mitchell Plateau and surrounding area.

STONEWORTS

CHAROPHYCEAE

CHARACEAE

Chara leptopitys A.Br.
Submerged aquatic; 'Caravan Creek' near Port Warrender.

Chara ? vulgaris (L.) R.D.W.
Submerged aquatic; Surveyors Pool Falls.

FUNGI

ASCOMYCETES

XYLARIACEAE

Daldinia concentrica Ces. & De Not.
On standing dead tree which had been burnt; on dead tree in vine thicket.

BASIDIOMYCETES

AURICULARIACEAE

Auricularia mesenterica Fr.
On bark of living *Ficus coronulata*; on dead wood in vine thicket.

GANODERMATACEAE

Ganoderma applanatum (Pers.) Pat.
On living *Eucalyptus tetrodonta.*

Ganoderma lucidum (Leys. ex Fr.) P. Karst.
On dead wood; in vine thicket on wood.

HYMENOCHAETACEAE

Phellinus gilvus (Schw.) Pat.
Growing on mangrove; on dead wood; on living *Melaleuca leucadendron.*

Phellinus robustus (P. Karst.) Bourd & Gatz.
On dead and fallen *Eucalyptus.*

PLEUROTACEAE

Panus fasciatus (Berk.) Pegler

Pleurotus sp. A.KFK 5164.
On dead wood.

Pleurotus sp. B.KFK 4905, 5292.
On dead, fallen *Livistona*; on dead wood.

POLYPORACEAE

Gloeophyllum concentricum G.H. Cunn.
On dead *Eucalyptus*.

Osmoporus sp. RJH/MP38.
Near ground in hole formed by upturned roots of *Livistona*.

Pseudofavolus tenuis (Hook.) G.H. Cunn.
On dead wood in vine thicket.

Pycnoporus coccineus/sanguineus KFK 4909, 4914, 5060, 5166; RJH/MP35/A; RJH/MP s.n.
On fallen dead *Livistona*; on fallen, dead *Eucalyptus*; on dead wood in vine thicket; on dead wood; on *Eucalyptus* near edge of swamp; on *Melaleuca leucadendron* in swamp.

Pycnoporus sp. KFK 4844.
Orange bracket fungi on fallen *Melaleuca*. Growing at Airfield Swamp.

Trametes lilacino-gilva (Berk.) Lloyd
On *Eucalyptus*.

Trametes muelleri Berk.
On dead wood in vine thicket; on fallen *Eucalyptus*.

Trichaptum cervino-gilvum (Jungh.) G.H. Cunn.
In vine thicket.

SCHIZOPHYLLACEAE

Schizophyllum commune Fr.
On dead wood.

SCLERODERMATACEAE

Pisolithus tinctorius (Mich. ex Pers.) Coker & Couch KFK 6715.
In lateritic earth on side of track.

THELOPHORACEAE

? Genus KFK 5347.
On dead wood.

GASTEROMYCETES

PODAXACEAE

Podaxis sp. KFK 7113, 7118.
On active termite mounds.

MYXOMYCETES

PHYSARACEAE

Physarum nutans Pers.
On rotting wood in vine thicket.

TRICHIACEAE

Arcyria cinerea (Bull.) Pers.
On rotting wood in vine thicket.

Hemitrichia serpula (Scop.) Rost.
On litter in vine thicket.

Perichaena depressa Libert.
On rotting bark on ground.

LICHENS

Buellia aff. *subdisciformis* RJH/MP1A.
On basalt in vine thicket.

Buellia sp. RJH/MP133.
On basalt in vine thicket.

Caloplaca sp. A. RJH/MP177D.
On smooth bark of tree in vine thicket.

Caloplaca sp. B. RJH/MP1B.

Catillaria sp. A. RJH/MP157B.
On basalt in vine thicket.

Coccocarpia sp. A. RJH/MP188.
On smooth bark of tree in vine thicket.

Cyanophytic lichens indet. RJH/MP141A; RJH/MP191C.
On basalt.

? Dermatocarpon sp. RJH/MP190C.
On lateritic rock.

Dirinaria aff. *applanata* RJH/MP157C.
On basalt in vine thicket.

Graphina aff. *sumplocorum* RJH/MP133B.

Graphina sp. A. RJH/MP139.
On tree bark near Walsh Point.

Graphis aff. *implicata* RJH/MP176A.
On smooth bark of tree in vine thicket.

Graphis sp. A. RJH/MP196B.
On smooth bark of *Randia cochinchinensis* in vine thicket.

Graphis sp. B. RJH/MP176A.
On smooth bark of tree in vine thicket.

Graphis sp. C. RJH/MP175E.
On smooth bark of tree in vine thicket.

? Gymnoderma sp. RJH/MP176B.
On smooth bark of tree in vine thicket.

? Heppia sp. RJH/MP129.
On exposed basalt near Walsh Point.

Icmadophila sp.
On basalt in vine thicket.

Lecanora sp. RJH/MP196A.
On smooth bark of *Randia cochinchinensis* in vine thicket.

Ochrolechia geminipara
On basalt SW of Walsh Point.

Ochrolechia aff. *pallescens* RJH/MP139.
On tree bark near Walsh Point.

Parmentaria microspora
On bark of tree near Walsh Point.

Parmentaria sp. RJH/MP177B.
On smooth bark of tree in vine thicket.

Pertusaria sp. RJH/MP37B, RJH/MP49.
On *Xanthostemon* near Camp Creek.

Physcia sp. RJH/MP157D.
On basalt in vine thicket.

Pleurotrema pyrenuloides
On smooth bark of tree in vine thicket.

Thelotrema sp. RJH/MP37B.
On *Xanthostemon* near Camp Creek.

Pyxine sp. RJH/MP37A.
On *Xanthostemon* near Camp Creek.

Gen. sp. indet. related to *Peltula* RJH/MP157B.
On basalt in vine thicket. Professor Dr A. Henssen, University of Marburg, W. Germany, considers this species to belong to an undescribed genus.

PTERIDOPHYTA (Ferns and Fern Allies). (Classification after Pichi Sermolli 1977).

ADIANTACEAE

†*Adiantum hispidulum* Sw.
Common fern growing in crevices of laterite at Camp Creek; below rim of lateritic breakaway of plateau.

Adiantum philippense L.
Growing on and near ledge under Mertens Falls.

BLECHNACEAE

Stenochlaena palustris (Burm. f.) Bedd.
Growing on sandstone, rhizome running along fissures.

GLEICHENIACEAE

Dicranopteris linearis (Burm. f.) Underwood
Growing on sandstone overhang, bottom of Surveyors Pool Falls near plunge pool.

LINDSAEACEAE

Lindsaea ensifolia Sw.
Growing on ledge under Surveyors Pool Falls; common.

LYCOPODIACEAE

Lycopodium cernuum L.
Erect fern ally, growing in seepage area near Surveyors Pool Falls.

NEPHROLEPIDACEAE

Nephrolepis hirsutula (Forst.) Pr.
Growing on sandstone below Mitchell Falls.

OPHIOGLOSSACEAE

Ophioglossum sp. (J. Smith s.n.)
Growing on basalt ledges beside creeks and in gorges.

PARKERIACEAE

Ceratopteris thalictroides (L.) Brongn.
Delicate fern. Growing on wet sandstone ledges under Mitchell Falls.

POLYPODIACEAE

Drynaria quercifolia (L.) J. Sm.
On lateritic breakaways above remnant vine thicket and on sandstone rock ledge; common.

SELAGINELLACEAE

Selaginella ciliaris (Retz.) Spring
Delicate ephemeral. Growing on wet sandstone ledges and in damp areas under lateritic boulders; common.

Selaginella sp. KFK 5303.
Growing in damp area under lateritic boulder at 'Black Mud Swamp'.

SINOPTERIDACEAE

Cheilanthes pumilio (R.Br.) Mett.
Small fern growing in grey/black silt on bank of Camp Creek under *Pandanus spiralis.*

Cheilanthes rotunda Bonap. or *C. contigua* Baker
Growing in monsoon forest near the crushing plant.

Cheilanthes sieberi Kuntze
Common on laterite at edge of palm/eucalypt woodland.

Cheilanthes tenuifolia (Burm.f.) Sw.
Growing along edge and in eucalypt woodland on laterite and on mudbanks beside creek.

Cheilanthes ? tenuifolia (Burm.f.) Sw. KFK 4899.
Dead fern on laterite in palm/eucalypt woodland.

Cheilanthes vellea (R.Br.) F. Muell.
Growing in rock crevices of lateritic plateau and in moist areas under the edge of lateritic boulders.

Cheilanthes sp. KFK 6966.
Woolly fern. Very common on shallow soil at falls east of basecamp.

Cheilanthes sp. KFK 6967.
Fern, growing on shallow soil apron surrounding exposed basalt sheets with KFK 6966.

Cheilanthes sp. KFK 6978.
Fern growing on red soil over laterite in palm/eucalypt woodland.

Cheilanthes sp. KFK 7120.
Fern growing in lateritic scree below breakaway rim of plateau.

Cheilanthes sp. KFK 7124.
Fern growing in black silt area which is seasonally wet.

GYMNOSPERMAE

CUPRESSACEAE

Callitris intratropica Baker & Smith
Small tree to 5 m. Growing on laterite with *Eucalyptus latifolia* and *Livistona eastonii.*

CYCADACEAE

Cycas basaltica C.A. Gardn.
Up to 4-5 m high. On basalt soils towards Port Warrender.

Cycas lane-poolei C.A. Gardn.
Up to 5.5 m high. Male has small cones. Male and female plants often branched. Growing on edge of wet swampy area, soil predominantly shallow over laterite.

MONOCOTYLEDONAE

AMARYLLIDACEAE

†*Crinum asiaticum* L.
Large bulbous plant with wide leaves, flowers white. Growing in Airfield Swamp where plants are seasonally inundated.

†*Crinum uniflorum* F. Muell.
Perennial bulbous plant with narrow leaves, flowers white. Growing in grey silt areas of plateau.

†*Eurycles alba* (R.Br.) Back
Bulbous lily, flowers white. Growing in dense patches in red clay over basalt.

APONOGETONACEAE

Aponogeton elongatus F. Muell.
Aquatic; inflorescence emergent; flowers yellow. Growing on edge of creek in basalt gorge.

ARACEAE

†*Amorphophallus galbra* F.M. Bail.
Bulbous annual growing in vine thicket south of mining company's trial crusher plant.

Colocasia antiquorum Schott
Tuberous aroid; spathe creamy-yellow. Growing in water of creek.

Typhonium liliifolium F. Muell. in Schott
Bulbous annual; leaves linear, light green with prominent veining and undulate margins; spathe green, surrounding ovary becoming deep burgundy. Growing in red soil over basalt on edge of plateau.

ARECACEAE

Livistona eastonii C.A. Gardn.
Palm, 7 m; fruits turning black, shining; very common on lateritic plateau.

COMMELINACEAE

Cartonema parviflorum Hassk.
Herb; flowers yellow. Growing in understorey of palm/eucalypt woodland.

Cartonema spicatum R.Br.
Bulbous herb. Growing in black soil at Camp Creek.

Commelina ensifolia R.Br.
Herb; flowers blue. Growing on lateritic plateau near costean.

Commelina sp. KFK 7012.
Prostrate glaucous herb. Growing in residual patch of vine thicket.

Commelina sp. KFK 7013.
Prostrate hairy herb. Growing on edge of remnant vine thicket.

Murdannia graminea (R.Br.) Bruckn.
Perennial herb; flowers white, on lateritic plateau.

CYPERACEAE

Cyperus aquatilis R.Br.
Sedge on banks of Camp Creek.

Cyperus breviculmis R.Br.
Sedge growing in wet areas of Airfield Swamp.

Cyperus conicus (R.Br.) Boeck.
Large sedge growing in shallow soil on edge of 'Caravan Creek'.

Cyperus difformis L.
Sedge growing on seepage area, Port Warrender.

Cyperus haspan L.
Tufted sedge on edge of Airfield Swamp.

Cyperus javanicus Houtt.
Large sedge growing on edge of creek at Port Warrender.

Cyperus spp. A.KFK 4873, RJH/MP 48, 52, 112, 113, 114, 115, 117, 118, 119.
Growing on edge of wet area.

Cyperus spp. KFK 5215, 6642.
Tufted caespitose sedge. Growing in seepage areas on laterite; at eroding edge of plateau; and forming dense swards behind mangroves at Port Warrender.

Eleocharis brassii S.T. Blake
Emergent aquatic, flowers white, feathery. Forming dense stands in Airfield Swamp.

Eleocharis caespitosissima Baker
Submerged sedge on edge of Airfield Swamp.

Fimbristylis arthrostyloides W.V. Fitzg.
Caespitose sedge. Growing on ledge under sandstone overhang.

Fimbristylis cephalophora F. Muell.
Caespitose sedge. Growing in 'Caravan Creek'.

Fimbristylis cymosa R.Br.
Sedge forming dense swards behind mangroves at Port Warrender.

Fimbristylis sp. KFK 4765, 4836, 7031, RJH/MP17, 27, 116.
Growing on edge of Airfield Swamp.

Fuirena ciliaris (L.) Roxb.
Sedge; stems and leaves yellowish/green. Growing in damp area beside 'Caravan Creek' and in closed grassland.

Fuirena umbellata Rottb.
Growing in 'Caravan Creek' amongst *Typha.*

Lipocarpha microcephala Kunth.
Beside 'Caravan Creek'.

Rhynchospora affinis W.V. Fitzg.
Common in closed grassland under *Eucalyptus apodophylla.*

Scleria aff. *pygmaea* KFK 4873, 4853.
Sedge on wet edge of Airfield Swamp.

Scleria sp. KFK 7024.
Sedge growing in closed grassland on black silt beside Camp Creek under *Eucalyptus papuana.*

DIOSCOREACEAE

Dioscorea bulbifera L.
Climber; flowers white; fruits narrowly winged. Growing on edge of plateau.

†*Dioscorea transversa* R.Br. vel. aff.
Climber; fruits broadly winged. Growing in vine thicket on edge of plateau.

ERIOCAULACEAE

Eriocaulon quinquangulare L.
Aquatic; flower heads silver in colour. Growing in running water of Camp Creek and common in wetter areas of Airfield Swamp.

Eriocaulon setaceum L.
Aquatic. Growing in Airfield Swamp.

Eriocaulon sp. KFK 4827; RJH/MP53.
Submerged aquatic; flower heads black.

FLAGELLARIACEAE

Flagellaria indica L.
Creeper; flowers cream. Growing on laterite; on sandstone with *Hoya* and in vine thicket.

HAEMODORACEAE

Haemodorum sp. KFK 6656.
Perennial; lower stem red; old seed capsules black.

HYDROCHARITACEAE

†*Vallisneria* aff. *caulescens* F.M. Bail. & F. Muell. KFK 4830.
Aquatic; lower leaves green becoming red; stem red; inflorescence scapes mainly red. Growing in 0.5 m of water in Airfield Swamp, common in areas of deeper water.

HYPOXIDACEAE

Curculigo ensifolia R.Br.
Tussocky herb; flowers yellow. Growing in black silt flats adjacent to Camp Creek.

JUNCAGINACEAE

Triglochin procera R.Br. var. *dubia* (R.Br.) Benth.
Rhizomatous perennial. In Mitchell River.

LILIACEAE

Asparagus racemosus Willd.
Climber; flowers green/white. Growing on volcanic, rocky loam on edge of vine thicket.

?*Chlorophytum laxum* R.Br. ASG 14499.
Herb with ephemeral leaves and inflorescences; flowers pale cream-green; fruit green.

Smilax australis R.Br.
Creeper. Growing with *Flagellaria* in 'Hidden Valley' and near Mertens Falls.

Tricoryne elatior R.Br.
Much branched tussock herb; flowers yellow. Growing in damp patch near Airfield Swamp.

NAJADACEAE

Najas graminea Delile
Submerged aquatic growing with *Vallisneria caulescens* in small pool adjoining Camp Creek.

ORCHIDACEAE

†*Calochilus holtzei* F. Muell.
Tuberous orchid growing under *Eucalyptus latifolia* on shallow soil over laterite.

†*Calochilus* sp. KFK 7110.
Orchid growing in seasonally inundated area under *Eucalyptus apodophylla* in closed grassland.

Cymbidium canaliculatum R.Br.
Epiphyte, flowers white. Growing on dead wood by Camp Creek.

Dendrobium dicuphum R.Br.
Epiphytic orchid, flowers white. Common on *Pouteria sericea* in vine thickets.

†*Dipodium punctatum* (J.E. Sm.) R.Br.
Flowers pale pink with purple labellum. Growing in laterite of palm/eucalypt woodland.

†*Geodorum densiflorum* (Lam.) Schlechter
Ephemeral orchid growing in black silt on edge of Camp Creek and in vine thicket.

†*Habenaria elongata* R.Br.
Flowers pure white. Growing among basalt rocks under *Terminalia.*

†*Habenaria ferdinandii* Schlechter.
Herb, usually associated with rotting wood. Growing in palm/eucalypt woodland and on black alluvial soil at Airfield Swamp.

PANDANACEAE

Pandanus aquaticus F. Muell. ex Warb.
Erect screw-pine growing beside Camp Creek, forming clumps.

Pandanus darwinensis var. *latifructus* (St. John) B.C. Stone
Screw-pine to 5 m. Growing on plateau in seasonally wet depressions.

Pandanus semiarmatus St. John
Screw-pine to 3 m, leaves green. Growing on alluvial soil beside seasonal watercourse off western edge of the Plateau, W. of airfield.

Pandanus spiralis R.Br.
Screw-pine to 5 m; spines readily break off on contact.

Pandanus spiralis R.Br. var. *convexus* (St. John) B.C. Stone
Screw-pine to 3 m; flowers cream. Common on plateau.

Pandanus spiralis R.Br. var. *multimammillatus* B.C. Stone
Screw-pine to 3 m, leaves glaucous. Growing on seasonally inundated black silt at headwaters to Camp Creek.

Pandanus spiralis R.Br. var. ? KFK 6783.
Screw-pine to 6 m; leaves slightly glaucous. Occurring in patches in wet alluvial valley.

Pandanus sp. KFK 6738.
Screw-pine to 8 m; leaves glaucous; fruits 30 cm in diameter.

PHILYDRACEAE

Philydrum lanuginosum Banks & Solander ex Gaertn.
Flowers yellow. Growing on moist creek banks and in dense stand in shallow water (depth 30 cm) of 'Black Mud Swamp'.

POACEAE

Arundinella nepalensis Trin.
Sporadic tall grass occurring in the central wetter areas of alluvial valley under *Pandanus* sp., *Eucalyptus apodophylla* and *Banksia dentata.*

Brachiaria holosericea (R.Br.) Hughes
Grass, inflorescence silver. Forms tussocks on shallow soil. Occasionally associated with termite mounds.

Brachiaria pubigera (R. & S.) S.T. Blake
Grass growing on ripped laterite of re-afforestation area near campsite.

Cenchrus elymoides F. Muell.
Caespitose perennial grass approx. 2.5 m high. Forming patchy distribution on laterite and forming part of tall closed grassland.

Coelorachis rottboellioides (R.Br.) A. Camus
Common caespitose grass; broadleaved. Growing in wet alluvial valley; in wet area of closed grassland and off lateritic plateau above remnant vine thicket.

Coelorachis sp. RJH/MP20.
Height 1.2 m. Growing inside open area of *Melaleuca* swamp under canopy of *Melaleuca.*

Cymbopogon bombycinus (R.Br.) Domin
Caespitose grass to 1.3 m; inflorescence silvery. Patchy in understorey in eucalypt/palm woodland. Growing near swamp on dry land.

Cymbopogon procerus (R.Br.) Domin
Caespitose perennial grass; common. Growing on exposed laterite on slope of plateau.

Digitaria bicornis (Lamk.) R. & S.
Grass growing outside Amax mess-hut in disturbed area.

Dimeria ornithopoda Trin.
Grass growing by edge of Mitchell River.

Dimeria sp. KFK 4923.
Growing on edge of running creek.

Echinochloa colonum (L.) Link
Caespitose grass near base camp.

Eleusine indica (L.) Gaertn.
Grass, growing in mown area behind Amax mess-hut.

Elytrophorus spicatus (Willd.) A. Camus
Flowers green.

Eragrostis bleeseri-cumingii complex ASG 14512.
Ephemeral grass. Growing in clay-loam with other grasses and occasional *Melaleuca viridiflora.*

Eragrostis japonica (Thunb.) Trin.
Caespitose grass forming large clumps; inflorescence maroon. Growing in red lateritic soil near Camp Creek.

Eragrostis aff. *tenax* Jedw. KFK 6594.
Grass growing in costean 3.5 m deep on lateritic plateau.

Eragrostis sp. RJH/MP89, MP92.
Grass near kitchens; Amax campsite.

Eriachne ciliata R.Br.
Grass with *Pseudopogonatherum contortum* and *Eriachne pauciflora.*

Eriachne festucacea F. Muell.
Grass growing in fast running water near Camp Creek.

Eriachne obtusa R.Br. var. *major* Ewart & Davies
Grass forming dense stand on disturbed area near airfield.

Eriachne pauciflora W.V. Fitzg.
Caespitose grass up to 1.5 m; semi-prostrate; rooting at nodes; height varies with soil depth. Growing with *Pseudopogonatherum contortum, Eucalyptus apodophylla, Pandanus* sp. and *Banksia dentata*; on mud mound in swamp; and at edge of Airfield Swamp.

Eriachne aff. *pauciflora* W.V. Fitzg. KFK s.n.
Common grass forming main understorey on exposed laterite under *Eucalyptus latifolia* low open woodland.

Eriachne sp. KFK 4882.
Caespitose grass. Growing on shallow soil area with laterite exposed at surface.

?*Eriachne* sp. KFK 6797.
Caespitose grass. Growing in crevices of exposed basalt with *Calandrinia uniflora.*

Eriochloa procera (Retz.) C.E. Hubb.
Grass. Growing in lawn outside kitchen of base camp.

Eulalia fulva (R.Br.) O. Kuntze
Height 2 m. Growing beside small intermittent creek in closed grassland with few emergent trees.

Germainia truncatiglumis (F. Muell. ex Benth.) Chumsri
Caespitose grass. Growing in Surveyor Creek (with water flowing around it) and alongside creek.

Heteropogon contortus (L.) Beauv. ex R. & S.
Grass to 1.5 m; common. Growing on lateritic scree near vine thicket; and in eucalypt/palm woodland.

Ischaemum sp. KFK 7030.
Grass forming stands to 1.5 m. Occasionally mixed with *Paspalum orbiculatum.* Growing at Airfield Swamp.

Iseilema filipes S.T. Blake

Panicum delicatum Hughes

Panicum mindanaense Herr.

Panicum trachyrhachis Benth.
Tall grass. Growing in deep water of Airfield Swamp.

Paspalum orbiculare Forst.f.

Caespitose grass. Growing in deeper water, beside fallen *Melaleuca* in Airfield Swamp and on edge of Camp Creek.

Pennisetum pedicellatum Trin.

Grass, 1.5 m high where receiving water. Common around campsite especially on deeper cultivated soil.

Plectrachne bynoei C.E. Hubb.

Caespitose grass. Growing on exposed lateritic ridge with very little soil and on laterite of re-afforestation area.

Plectrachne melvillei C.E. Hubb.

Caespitose grass, common on sandstone ledge surrounding falls.

Plectrachne pungens (R.Br.) C.E. Hubb. var. *callosum* C.E. Hubb.

Grass in tussocks 0.75 m in diameter, 0.4 m high. Growing at edge of Airfield Swamp on laterite and on dry ground of plateau near swamp.

Pseudopogonatherum contortum (Brongn.) A. Camus

Caespitose, ephemeral grass, 2-20 cm in diameter; flowers brown; inflorescence black. Forming dense patches in wet alluvial valley and on shallow lateritic soil area (consisting mainly of this grass and *Eriachne* sp.); on low grassland; on lateritic rocks on plateau; and near eucalypt/palm woodland.

?*Rottboellia* sp. KFK 7128.

Grass to 1.5 m, dark green. Growing on silt area which is seasonally wet. (Not either of the two known Australian species: M. Lazarides, Herbarium Australiense — pers. comm.).

Sacciolepis indica (L.) Chase

Caespitose grass. Common in gravelly mud fringing Camp Creek.

Sacciolepis myosuroides (R.Br.) A. Camus

Grass on edge of woodland.

Schizachyrium pachyarthron C.A. Gardn.

Ephemeral grass.

Schizachyrium sp. KFK 4787.

Grass, dead and dry.

Setaria apiculata (Scribu. & Merr.) K. Schum.

Growing on edge of Airfield Swamp.

Sorghum australiense Garber & Snyder

Caespitose grass to 2.5 m; inflorescence golden hairy. One of many species forming tall closed grassland with few eucalypts emerging.

Sorghum aff. *brevicallosum* Garber KFK 6576, 6650.

Grass to 2.5 m. Growing on damp area beside track and outside vine thicket, forming dense patches.

Sorghum aff. *brevicallosum* Garber KFK 6636.

Grass forming dense belt in outer, wet perimeter of Airfield Swamp.

Sorghum plumosum (R.Br.) Beauv.

Tall, many stemmed grass growing near Camp Creek and in wet alluvial valley.

Sorghum sp. A.KFK 6781; RJH/MP84, MP91.

Grass, forming dense stand on outer perimeter of wet alluvial valley and near vine thicket.

Sorghum sp. B.KFK 4753, 4767, 5205, 5167; RJH/MP1, MP58.

Caespitose grass to 2.5 m; golden tomentum. Growing on edge of Airfield Swamp on laterite and in dense stands forming closed grassland in wet areas of swamp.

Sporobolus virginicus (L.) Kunth.

Tussocky grass. Growing on saline mud flat between mangroves and coastline.

Stipa sp. KFK 7044.

Caespitose grass. Growing on sandstone ledge above Surveyors Pool Falls.

Themeda arguens (L.) Hack.
Common grass. Forms understorey on laterite with *Eucalyptus miniata, E. tetrodonta, Livistona eastonii* woodland and at top of ridge near vine thicket.

Themeda australis (R.Br.) Stapf.
Caespitose grass. Forms dense stands near wet area and on laterite; on laterite under palms and beside small, intermittent creek in closed grassland with a few emergent trees.

Triodia pungens R.Br.
Viscid grass. Growing in large clumps amongst sandstone rocks.

Triodia sp. KFK 5047.
Common grass. On sandstone below waterfalls.

TACCACEAE

Tacca leontopetaloides (L.) Kuntze
Bulbous perennial, leaf segments broad. Common on plateau in palm/eucalypt woodland.

Tacca maculata Seem.
Bulbous perennial, leaf segments linear. Common on plateau in palm/eucalypt woodland.

TYPHACEAE

Typha domingensis Pers.
Clumps to 2.5 m. Growing in 'Caravan Creek'.

XANTHORRHOEACEAE

Lomandra tropica A. Lee ssp. *tropica.*
Tufted herb, flowers yellow. Growing in red soil under *Livistona* and *Eucalyptus latifolia.*

XYRIDACEAE

Xyris complanata R.Br.

Xyris aff. *complanata* R.Br. KFK 4851, 4860.
Flowers yellow. Growing in grassland.

Xyris aff. *elongata* Rudge KFK 4945.
Caespitose herb; flowers yellow. Growing in fast running Camp Creek and on edge of closed grassland of Airfield Swamp.

Xyris aff. *indica* L. KFK 6721.
Leaves reddish at base; flowers yellow; scape green.

Xyris aff. *oligantha* Steud. KFK 6588, 6819.
Herb; stems green; leaves tinged red/maroon, flowers yellow. Growing on scraped laterite/gravel area in grey silt.

Xyris aff. *pauciflora* Willd. KFK 4946.
Growing on flat area near Camp Creek.

Xyris sp. KFK 6582.
Basal leaves tinged red; stems green; flowers yellow. Growing in wet perimeter of Airfield Swamp with *Utricularia.*

DICOTYLEDONAE

ACANTHACEAE

Dicliptera glabra Dcne.
Dense bush. On disturbed laterite near costean.

Dicliptera aff. *leonotis* Dalz. ex Clarke KFK 5154.
Small herb; flowers pink/purple. Near edge of vine thicket.

Dicliptera aff. *leonotis* Dalz. ex Clarke KFK 6653.
Shrub up to 50 cm; flowers red/magenta.

Hypoestes floribunda R.Br.
Spindly shrub/herb; flowers pale cream/yellow suffused purple. Growing on area near saline mangrove flats, near Camp Creek and in eucalypt/palm woodland.

Hypoestes suaveolens C.A. Gardn.
Herb to 45 cm; flowers mauve. In vine thicket on coast.

Thunbergia ? arnhemica F. Muell. ASG 14490; KFK 6990.
Climbing perennial creeper, flowers white. Common in rocky loam on edge of vine thicket.

AMARANTHACEAE

Achyranthes aspera L.
Growing in vine thicket amongst scree.

Gomphrena canescens R.Br.
Herb; bracts bright pink. Growing in red soil over laterite.

Gomphrena flaccida R.Br.
Herb to 30 cm; flowers pink. Common in and on edge of woodland and on sandstone near Surveyors Pool Falls.

Gomphrena ? leptoclada Benth. ASG 14486.

Gomphrena sp. A.KFK 4813, 6607.
Herb; outer tepals dark red, inner orange. Growing on laterite.

Gomphrena sp. B.KFK 6762, 6608.
Small herb; flowers deep pink. Growing on laterite.

Gomphrena sp. C.KFK 5200, 6667; D.E. Symon 10236.
Erect herb; flowers white; not common. Growing on plateau in closed grassland and on coastal strip behind mangroves.

Ptilotus conicus R.Br.
Erect ephemeral herb; flowers deep magenta. Common on plateau in closed grassland and on laterite near Camp Creek.

Ptilotus polystachyus (Gaud.) F. Muell. f. *rubriflorus* (J.M. Black) Benth.
Erect herb growing in sandstone crevices at gauging station on Camp Creek.

Pupalia lappacea (L.) Juss.
Herb with clinging seeds. In vine thicket.

ANACARDIACEAE

Buchanania obovata Engl.
Tree/shrub up to 3 m; sterile. Growing on laterite and on edge of woodland near Airfield Swamp.

ANNONACEAE

†?*Fissistigma* sp. KFK 7096.
Straggly shrub to 4 m, leaves dark green. This species forms the main understorey to 'Crusher Vine Thicket'.

APIACEAE

Trachymene didiscoides (F. Muell.) B.L. Burtt
Erect shrub/herb. Growing on sandstone.

Trachymene sp. KFK 4792.
Erect herb; on sandstone.

Trachymene sp. KFK 6617.
Herb. Growing in disturbed laterite adjacent to mining costean.

APOCYNACEAE

Alstonia actinophylla (A. Cunn.) K. Schum.
Shrub to 2 m; leaves shiny, growing on edge of Camp Creek.

†*Alyxia spicata* R.Br.
Tree to 4 m; flowers white; fruit yellow/green.

Ervatamia orientalis (R.Br.) Domin
Tree/shrub to 2 m growing on sandstone outcrops at Surveyors Pool Falls and in shade of *Pouteria sericea.*

Parsonsia velutina R.Br.
Climber on trees. Growing in vine thicket.

Parsonsia ? velutina R.Br. KFK 4800.
Creeper on trees at Camp Creek.

Sarcostemma australe R.Br.
Scrambler on bushes in vine thicket.

Wrightia pubescens R.Br.
Deciduous tree to 8 m; long pods; white latex. In vine thicket.

ASCLEPIADACEAE

Cynanchum ? floribundum R.Br. ASG 13156.
Prostrate creeper; fruit present; sap milky.

Hoya australis R.Br.
Growing on rocks above Surveyors Pool Falls.

Tylophora flexuosa R.Br.
Climber to 1.2 m on tree. Growing on edge of lateritic plateau.

Tylophora sp. KFK 7000.
Creeper growing on basalt ledges and at Mining Camp.

ASTERACEAE

Blumea integrifolia DC.
Herb; radially creeping; flowers yellow. Growing in damp patch on laterite and regenerating near Surveyors Pool Falls.

Blumea mollis (D. Don) Merr.
Herb. Near edge of track to Port Warrender under *Pandanus* in water seepage area.

†*Blumea saxatilis* Zoll. & Mor.
Ephemeral herb; leaves bluish green; flowers yellow. Growing on wet soil by Camp Creek.

Blumea sp. KFK 4759, 4763.
Herb. Growing on bases of fallen *Melaleuca* in Airfield Swamp and on damp edge of swamp.

Chrysogonum ecliptoides (F. Muell.) Benth. and Hook. ex F. Muell.
Flowers yellow. Growing amongst grassland outside vine thicket and on laterite in water-shed area.

Chrysogonum ? ecliptoides (F. Muell.) Benth. and Hook. ex F. Muell. KFK 6665/A.
Flowers deep orange; common. Growing amongst grass on plateau.

Emilea sonchifolia (L.) DC.
Decumbent herb; flowers mauve. In disturbed ground adjacent to Amax basecamp; probably introduced.

Epaltes australis Less.
Small herb. Growing in disturbed area surrounding Mining Camp.

†*Minuria* sp. nov. (ined.) RJH/MP103.
Perennial herb, florets white. Growing in palm/eucalypt woodland.

Pleurocarpaea denticulata Benth.
Sub-shrub to 0.5 m, flowers white occasionally tinged maroon. On laterite in eucalypt/palm woodland.

Tridax procumbens L.
Spreading herb; ray florets white-cream. Common around Mining Camp.

Vernonia cinerea Less.
Flowers mauve/purple. On shallow soil on edge of lateritic plateau above vine thicket; on disturbed soil surrounding Mining Camp; probably introduced.

Vittadinia sp. KFK 6803.
Herb; flowers mauve.

Wedelia asperrima (Dcne.) Benth.
Erect herb growing on red laterite amongst closed grassland.

BIGNONIACEAE

Dolichandrone heterophylla (R.Br.) F. Muell.
Tree up to 6 m; leaves dull green; pods light brown. Growing on edge of vine thicket.

BOMBACACEAE

Adansonia gregorii F. Muell.
Baobab to 5 m; few leaves; few mature fruits. On basalt soils, Port Warrender.

Bombax ceiba L. var. *leiocarpum* A. Robyns
Deciduous tree to 15 m; bark smooth and light grey with peeling flakes up to 2 m from ground; leaves palmate; flowers red. Common emergent of vine-thicket.

Camptostemon schultzii Masters
Mangrove to 3 m; silvery foliage, Port Warrender.

BORAGINACEAE

Heliotropium strigosum Willd.
Herb. On damp areas.

Heliotropium sp. KFK 6788.
Semi-woody herb; flowers white. Growing on Carson Volcanics.

BURSERACEAE

Canarium australianum F. Muell.
Tree up to 15 m (3-15 m), well developed trunk; bark silver/grey and finely flaky; leaves dark green. Growing in and along edges of vine thickets and on sandstone below Surveyors Pool Falls.

†*Garuga floribunda* Dcne.
Tree up to 15 m; bark rough, light grey/brown; foliage occasionally tinged red giving the appearance of *Rhus*; fruits black/purple in colour. Forms main canopy in vine thickets and fringing riverine communities.

BYBLIDACEAE

Byblis liniflora Salisb.
Flowers lilac. Growing on edge of swamps.

CAESALPINIACEAE

Cassia absus L.
Sub-shrub to 0.5 m, flowers yellow. Growing on laterite in eucalypt/palm woodland.

Cassia glauca Burm.f.
Open branched shrub to 3 m; flowers yellow; pods dark. On edge of vine thicket and extending into thicket.

Cassia mimosoides L.
Small herb; flowers yellow/orange. Growing in grassland and at edge of Airfield Swamp.

Cassia sp. KFK 5192.
Shrub to 2 m; flowers yellow.

Erythrophleum chlorostachys (F. Muell.) Baill.
Erect tree to 5 m. Growing beside Camp Creek.

Lysiphyllum cunninghamii (Benth.) De Wit.
Tree up to 9 m. On wet raised area in swamps; on landward edge of mangroves; on edge of vine thickets.

Piliostigma malabaricum (Roxb.) Benth.
Tree to 5 m; bark brown and slightly fissured, flowers white. On raised grassy area of alluvial valley with *Pandanus.*

CAPPARACEAE

Cadaba capparoides DC.
Sprawling shrub up to 2 m; flowers white; fruits green; common in vine thickets.

Capparis quiniflora DC.
Climber 8 m above ground; flowers white and scented.

Capparis umbonata Lindl.
Tree to 3 m, flowers white. Growing on basalt above Port Warrender.

Capparis sp. A.KFK 5107, 5231, 6677.
Thorny climber. In vine thickets.

?*Capparis* sp. B.KFK 5343.
Climber, clinging by thorns. In vine thicket.

Cleome tetrandra DC.
Herb, flowers yellow. Growing on exposed basalt on E. side of Camp Creek.

Justago cleomoides (F. Muell.) O. Kuntze
Shrub to 40 cm, flowers yellow. Growing in shallow soil between massive sandstone outcrops S.W. of camp.

CARYOPHYLLACEAE

Polycarpaea corymbosa (L.) Lam.
Herb; stems red. Growing in shallow soil over laterite or in crevices in laterite of plateau.

Polycarpaea longiflora F. Muell.
Erect herb; flowers red. Growing on sandstone near Mining Camp and Mertens Falls.

CELASTRACEAE

Cassine melanocarpa (F. Muell.) O. Kuntze
Shrub/tree up to 5 m; fruits deep green, shiny and hard. In vine thicket.

Denhamia obscura (A. Rich.) Meisn.
Tree to 7 m; bark grey and smooth. Forms uppermost emergent in vine thicket and growing on sandstone around Mitchell Falls area.

†*Maytenus* aff. *emarginatus* (Willd.) Ding Hou KFK 7129.
Trees to 4 m growing on sandstone S.W. of Camp Creek.

CHENOPODIACEAE

Suaeda australis (R.Br.) Moq.
Diffuse shrub. Growing on raised ridges on landward side of mangrove flat and on saline flat in front of *Avicennia*, Port Warrender.

CLUSIACEAE

Calophyllum sil Laut.
Tree to 14-15 m, leaves shiny green. Growing on exposed basalt at edge of Camp Creek and in vine forest.

Hypericum gramineum Forst.f.
Herb; stems tinged maroon; flowers yellow/orange. Growing on edge of Camp Creek.

COCHLOSPERMACEAE

Cochlospermum fraseri Planch.
Shrub/tree up to 2.5 m; flowers yellow. Growing on laterite and sandstone.

COMBRETACEAE

Lumnitzera racemosa Willd.
Mangrove to 2 m, flowers white. Growing in 'Caravan Creek'.

Terminalia canescens (DC.) Radlk.
Straggly shrub/tree to 6 m (3-6 m); bark dark brown and deeply fissured. Common on Carson Volcanics and on shallow soil areas of lateritic plateau, as well as on open lateritic areas outside of *Eucalyptus miniata-E. tetrodonta* woodland.

Terminalia carpentariae C.T. White
Tree to 6 m, bark grey, leaves glaucous in appearance due to hairs. Growing on massive outcrop of sandstone at falls.

Terminalia cunninghamii C.A. Gardn.
Tree to 5 m, bark dark. Fruits large and green. Growing on sandstone E. of Surveyors Pool Falls.

Terminalia ferdinandiana Exell
Tree to 5 m, bark dark. Growing on sandstone S.W. of Camp Creek.

Terminalia fitzgeraldii C.A. Gardn.
Trees to 10 m, bark brown and deeply fissured; fruits large, globular, occasionally beaked, purplish black at maturity. Forming dense stands on alluvial flats with *T. canescens*; on Carson Volcanics and alongside creeks in alluvial valleys.

Terminalia hadleyana W.V. Fitzg.
Tree up to 8 m (3-8 m). Growing on sandstone and laterite.

Terminalia petiolaris A. Cunn. ex Benth.
Tree up to 14 m (5-14 m); bark dark brown with deep anastomosing ridges; fruit green turning black/purple. Common in vine thickets; directly behind mangroves on rocky headland and along fresh-water creek.

Terminalia platyphylla F. Muell.
Tree to 12 m; well developed canopy. Growing by creek.

CONVOLVULACEAE

Evolvulus alsinoides L.
Low herb, flowers blue. Growing on exposed Carson Volcanics.

Evolvulus sp. KFK 5243.
Low herb, flowers pale blue. On edge of vine thicket.

Ipomoea brasiliensis (L.) Sw.
Prostrate creeper. Recorded from Admiralty Gulf by C.A. Gardner.

Ipomoea gracilis R.Br.
Scrambler, flowers white. Common on plateau.

Ipomoea graminea R.Br.
Prostrate herb with erect leaves.

Ipomoea sp. KFK 6643.
Scandent shrub; reddish stems; seeds black with rugose elaiosome. Growing on edge of Camp Creek.

Merremia aff. *quinata* (R.Br.) Oostr. RJH/MP110.
Creeper on vegetation growing beside 'Caravan Creek'.

Merremia tridentata (L.) Hall.f. subsp. *hastata* (Hall.f.) Oostr.
Creeper on vegetation growing behind tidal mud flat at Port Warrender.

Operculina sp. KFK 5123.
Climber growing on trees in vine thicket.

CUCURBITACEAE

Diplocyclos palmatus (L.) C. Jeffrey
Growing on trees in vine thicket and on rocks next to 'Caravan Creek'.

Luffa graveolens Roxb.
Creeper. Growing on sandstone.

Trichosanthes sp. KFK 7010.
Climber growing in remnant patch of vine thicket at Port Warrender.

DILLENIACEAE

Hibbertia sp. A.KFK 6739.
Much branched shrub to 40 cm; flowers yellow. On lateritic scree slope.

Hibbertia sp. B.KFK 6807.
Growing on ripped laterite of re-afforestation area.

Hibbertia sp. KFK 7048, 7069.
Shrub, flowers yellow. Growing in palm/eucalypt woodland.

Hibbertia sp. RJH/MP47.
Growing in palm/eucalypt woodland.

Pachynema junceum Benth.
Perennial herb; basal stems reddish; petals red, sepals pinkish-brown, margins white. Growing on plateau in palm/eucalypt woodland.

DROSERACEAE

Drosera burmannii Vahl.
Leaves green or red; flowers red/brown. Growing on damp patch at edge of Airfield Swamp.

Drosera indica L.
Herb; flowers pink. Growing in closed grassland in wet area under *Eucalyptus apodophylla*, and on edge of Airfield Swamp.

Drosera petiolaris R.Br. ex DC.
Herb; flowers white/creamy yellow. On red clay and wet areas of Airfield Swamp.

Drosera sp. KFK 7059.
Rosetted herb, flowers cerise. Common on areas of shallow soil amongst massive sandstone, E. of Surveyors Pool Falls.

EBENACEAE

Diospyros ferrea (Willd.) Back. var. *humilis* (R.Br.) Back.
Tree to 12 m; bark flaky. Growing in vine thicket.

Diospyros ? montana Roxb. RJH/MP77.
Shrub growing in vine thicket E. of Surveyors Pool Falls.

Diospyros nitens W.V. Fitzg.
Tree to 7 m; fruits green. Growing in vine thicket.

ERYTHROXYLACEAE

Erythroxylum ellipticum R.Br.
Erect shrub/small tree. Growing among sandstone rocks near Surveyors Pool Falls.

EUPHORBIACEAE

Antidesma ghaesembilla Gaertn.
Tree/shrub to 5 m; bark brown; leaves shiny green; fruits green, sometimes tinged maroon, and fleshy. Growing on raised ridges in wet alluvial valley and beside Camp Creek.

Croton sp. KFK 5100B.
Tree in vine thicket.

Drypetes australasica (J. Muell.) Pax Hoffm.
Small tree to 4 m, leaves shiny green, fruits yellow turning bright red. Growing on edge of vine thicket amongst lateritic scree.

Euphorbia hirta L.
On red soil near Camp Creek; probably introduced.

Euphorbia plumerioides Hassk.
Shrub to 3 m; deciduous; flowers small, green. In vine thicket on the coast.

Euphorbia vachellii Hook. & Arn.
Flowers white. On moist area near Camp Creek.

Euphorbia spp. KFK 4841, 4911, 4921, 5059, 5185, 5194, 5254, 5303/A, 6704, 6775; RJH/MP97.

Excoecaria agallocha L.
Mangrove to 3 m. On landward edge of mangal, Port Warrender.

Glochidion disparipes Airy Shaw
Shrub to 5 m; leaves dark green; fruit fleshy pink with white markings. On edge of woodland.

Petalostigma pubescens Domin
Dense shrub/tree to 3 m; fruits orange. Common on sandstone and in eucalypt/palm woodland.

Phyllanthus aff. *polycladus* W.V. Fitzg. KFK 5056.
Growing on sandstone near Mertens Falls.

Phyllanthus sp. KFK 4786, 5321, 6711, 6821.
Openly branched shrub to 3 m; leaves pale green; fruits green and fleshy turning black. Growing on wet area beside Camp Creek and on edge of 'Black Mud Swamp'.

Ricinocarpos cf. *marginatus* Benth. Maconochie 1277.
Erect shrub to 2 m; blue grey in colour; fruits at apex of shoot. Growing amongst sandstone rocks, Surveyors Pool Falls.

Ricinocarpos sp. KFK 5007.
Shrub. Growing on sandstone.

Sebastiana chamelaea (L.) Muell. Arg.
Erect herb. Stock ? perennial.

GENTIANACEAE

Canscora diffusa R.Br.
Herb; flowers pink. Growing by creek, on lateritic rocks in creek and in wet area under Mertens Falls.

Nymphoides aff. *parvifolia* (Griseb.) Kuntze KFK 4747, 4760.
Semi-aquatic; flowers white; leaves crescent shaped. On edge of Airfield Swamp.

Nymphoides aff. *hydrocharoides* (F. Muell.) Kuntze KFK 4758, 4826.
Aquatic; flowers orange. Growing in pools of water in Airfield Swamp.

Nymphoides sp. KFK 6589.
Floating aquatic, flowers white, seeds smooth, shiny and black. Growing in Airfield Swamp.

GOODENIACEAE

Calogyne holtzeana Specht
Flowers yellow. Growing on lateritic scree.

Goodenia propinqua W.V. Fitzg.
Basally rosetted; flowers yellow. Growing in damp patch.

Goodenia pumilio R.Br.
Plant growing in clumps; flowers deep burgundy in colour. Growing on black silt under *Eucalyptus apodophylla* in seasonally wet area.

Goodenia sepalosa F. Muell. ex Benth.
Semi-woody herb; flowers yellow. In eucalypt/palm woodland.

Goodenia sp. KFK 6727.
Upright to sprawling herb; flowers maroon with darker maroon centres. Growing in wet areas.

Goodenia sp. KFK 7136/A, 7137, 7134, 7130.
Prostrate to ascending herb; flowers yellow, growing in shallow pockets of soil between massive sandstone outcrops S.W. of Mining Camp and near creek.

HALORAGACEAE

Gonocarpus leptothecus (F. Muell.) Orchard
Growing in wet area by Camp Creek.

Myriophyllum trachycarpum F. Muell.
Growing on edge of Camp Creek.

LAMIACEAE

Dysophylla verticillata Benth.
Aquatic; stem tinged red; flowers pink/mauve. Growing in Airfield Swamp.

†*Hyptis suaveolens* (L.) Poit.
Growing around Amax campsite, probably introduced.

LAURACEAE

Cassytha sp. A.KFK 5118.
Parasitic vine. Growing on burnt sandstone area.

Cassytha sp. B.KFK 6768, 6788/A.
Flowers white. Growing on plateau amongst grasses.

Cassytha sp. C.RJH/MP67.
Growing on several species of shrub near Mitchell Falls.

Cryptocarya sp. KFK 5028.
Tree to 15 m. Growing directly below Mertens Falls.

Litsea sp. KFK 5145, 6971.
Tree to 12 m. Growing in vine thicket; on basalt below falls.

LECYTHIDACEAE

Planchonia careya (F. Muell.) R. Knuth
Tree/shrub up to 16 m (2-16 m); bark deeply fissured; foliage dense, sometimes tinged red, flowers white. Growing on laterite; on sandstone near Camp Creek; on the edge of vine thicket and eucalypt woodland.

LENTIBULARIACEAE

Utricularia chrysantha R.Br.
Herb; lower stem reddish, upper green; flowers yellow; common. Growing in moist areas on edge of swamp; at foot of Surveyors Pool Falls; and in mud at edge of Camp Creek.

Utricularia exoleta R.Br.
Herb; bladders either free floating or rooted; flowers yellow. Growing in waters of Camp Creek.

Utricularia kimberleyensis C.A. Gardn.
Herb; flowers purple, throat yellow, reverse side of corolla light yellow. Growing on edge of swamp.

Utricularia lasiocaulis F. Muell.
Herb up to 10 cm; flowers purple with yellow/white inside throat, outside yellow/purple with deeper striations, petals bronze on back. Growing on mud at edge of Airfield Swamp.

†*Utricularia leptoplectra* F. Muell.
Herb; flowers pale mauve, centre of throat white surrounded by deep purple; common. In wet area by swamp.

Utricularia limosa R.Br.
Diminutive erect herb; flowers white/cream. Growing in mud and deeper areas of Airfield Swamp.

†*Utricularia muelleri* Kam.
Floating aquatic; bladders and leaves mainly red; flowers whitish yellow tinged red.

Utricularia sp. KFK 6586.
Flowers white with mauve tinge on end of spur. Growing near Airfield Swamp.

Utricularia sp. KFK 7060.
Herb, flowers mauve. Growing on seepage area below Surveyors Pool Falls.

Utricularia sp. KFK 7115.
Bladderwort growing in mud of sandstone creek.

Utricularia sp. RJH/MP5.
Growing on edge of swamp on laterite.

LOBELIACEAE

Isotoma armstrongii E. Wimm.
Herb; flowers blue with white throat. In swampy closed grassland.

Lobelia dioica R.Br.
Slightly creeping herb; flowers blue. In Airfield Swamp ca. 6 m from waterline.

LOGANIACEAE

Mitrasacme connata R.Br.
Herb; flowers yellow, edge of perianth lobes slightly inflexed and bronze in colour.

Mitrasacme elata R.Br.
Stems reddish in lower half and leafless, upper stems green and leafless; flowers yellow, paler outside.

Mitrasacme foliosa C.A. Gardn.
Foliose plant. On mud flat near camp.

Mitrasacme gentianea F. Muell.
Herb; stems tinged orange-pink; flowers white, tinged orange-pink in bud; strong curry odour.

Mitrasacme laevis Benth.
Herb; flowers white. Growing in wet area of swamp.

Mitrasacme aff. *lepidocalyx* W.V. Fitzg. KFK 5354.
Flower white, tinged orange.

Mitrasacme nudicaulis Benth.
Basally rosetted; flowers white. Growing in wet grassland.

Mitrasacme sp. KFK 5349, 6736, 6982, 7147.
Leafy stemmed species; flowers white.

Strychnos lucida R.Br.
Shrub to 3 m; fruit yellow/green and papery, containing sticky seeds. Growing on sandstone.

LORANTHACEAE

Amyema benthamii (Benth.) Dans.
Hemi-parasite. Flowers orange/green. In vine thickets.

Amyema bifurcatum (Benth.) Tiegh.
Growing on *Eucalyptus* sp. (KFK 4971).

Amyema sanguineum (F. Muell.) Dans.
Hemi-parasite on *Eucalyptus tetrodonta*; perianth pink/red.

Amyema thalassium Barlow
Hemi-parasite on *Avicennia marina* in mangrove belt; leaves large and fleshy.

Decaisnina signata (F. Muell. ex Benth.) Tiegh.
Hemi-parasite on *Buchanania* and *Eucalyptus*; flowers yellow to orange/red.

LYTHRACEAE

Ammannia indica Lam.
Flowers dark red. At edge of tidal mud flats, behind mangrove belt.

Ammannia multiflora Roxb.
Small herb; stems and fruits red. Growing in 'Caravan Creek'.

†*Lagerstroemia archerana* F.M. Bail.
Tree to 15 m, flowers purple. Growing in vine thicket.

Lythrum arnhemicum F. Muell.
Herb with fleshy base, flowers pink. On edge of Airfield Swamp.

Rotala verticillata L.
Small herb. On mud near Airfield Swamp.

Rotala sp. KFK 4858.
Herb. In closed swampy grassland, Airfield Swamp.

MALVACEAE

Abelmoschus ficulneus (L.) Wight and Arn.
Erect herb to 1.5 m; annual; flowers white to pink with dark centre. Growing in heavy black soil and clay soil of dry swamp.

Abelmoschus moschatus Medik. ? ssp. *biakensis* (Hocht.) Borss.
Prostrate herb; stems maroon; flowers yellow. On lateritic plateau.

Abelmoschus moschatus Medik. ssp. *tuberosus* (Span.) Borss.
Scrambling herb, flowers white with deep maroon centre. On black silt.

Gossypium pilosum Fryxell
Scrambling herb; flowers white to pink with deep maroon centres and yellow staminal column. Growing on laterite, Carson Volcanics and sand near creek.

Gossypium ? pilosum Fryxell ASG 13155.
Prostrate creeper; flowers white or pink.

Gossypium populifolium (Benth.) F. Muell.
Prostrate creeping herb; flowers pink-white with red centre. Growing on lateritic plateau.

Hibiscus leptocladus Benth.
Shrub to 1 m; flowers blue. Growing in sandstone.

Hibiscus superbus C.A. Gardn.
Shrub to 2.5 m, flowers pink with deep maroon centre. Common on sandstone.

Thespesia populneoides (Roxb.) Kostel.
Tree to 3 m. Growing on strand behind mangroves at Port Warrender.

MELASTOMATACEAE

Osbeckia australiana Naud.
Shrub to 1 m; flowers purple. Growing on edge of Camp Creek.

MELIACEAE

Aglaia sp. KFK 5099, 5131, 5234.
Tree up to 12 m (5-12 m); buds mustard in colour. In vine thicket.

Dysoxylum sp. KFK 5134, 6963.
Tree 5-8 m, growing in vine thicket near Camp Creek and edge of riverine forest at falls.

Owenia vernicosa F. Muell.
Tree to 5 m; common. Growing on sandstone.

Xylocarpus australasicus Ridley
Mangrove to 3 m; shiny green leaves; flowers whitish/cream. On landward edge of mangal, Port Warrender.

MENISPERMACEAE

†*Pachygone ovata* Miers ex Hook.f. & Thomas vel. aff.
Common climber in vine thickets.

MIMOSACEAE

Acacia bidwillii Benth.
Erect shrub/tree to 3-4 m. Growing by creek, along coastline and on Carson Volcanics.

Acacia delibrata (A. Cunn.) Benth.
Tree to 5 m; mini-ritchie bark; sterile.

Acacia gardneri Maiden & Blakely
Tree to 3 m; bark glaucous; flowers yellow. On sandstone plateau.

Acacia holosericea A. Cunn.
Tree to 3 m. On sandstone.

Acacia kelleri F. Muell.
Spreading bush to 1 m. On sandstone near and away from Camp Creek.

Acacia orthocarpa F. Muell.
Compact shrub to 2 m. Beside creek on sandstone.

Acacia pallidifolia Tindale
Tree to 5 m; bark corky and fissured; immature pods green, mature pods light brown. Growing on Carson Volcanics of plateau slope.

Acacia platycarpa F. Muell. Marshall s.n.

Acacia retinervis Benth.
Openly branched tree/tall shrub to 5 m; bark dark brown and fissured; flowers yellow and faintly fragrant. Growing on shallow lateritic gravel over Carson Volcanics.

Acacia stigmatophylla A. Cunn.
Shrub to 1.2 m. Growing on Carson Volcanics, Port Warrender.

Acacia suberosa (A. Cunn.) Benth.
Shrub to 2 m, bark corky. Growing on shallow laterite N.E. of Surveyors Pool Falls.

Acacia translucens A. Cunn. ex Hook.
Shrub 1-1.5 m. The most common species regenerating on grid line scrapes; common on sandstone.

Acacia tumida F. Muell. ex Benth.
Much branched shrub to 3 m, pods brown. Growing in shallow soil on plateau.

Acacia sp. A.KFK 6659.
Much branched shrub to 1.5 m. On shallow soil over laterite in *Eucalyptus tetrodonta-E. nesophila* woodland.

Acacia sp. B.KFK 6750.
Shrub to 2 m, flowers pale yellow and fragrant. Growing on shallow laterite with *Callitris intratropica.*

Acacia sp. C.KFK 6751.
Shrub to 2.5 m, leaves glaucous, flowers yellow and fragrant. Growing on shallow laterite with *Callitris intratropica.*

Acacia sp. D.KFK 6799.
Openly branched shrub to 2.5 m, pods brown. On shallow soil over basalt near Mining Camp.

Albizia lebbek (Willd.) Benth.
Tree 12-15 m; bark strongly lenticellular; leaves pinnate; pods whitish. Growing near edge of Camp Creek and in vine thickets.

MORACEAE

Ficus coronulata Miq.
Tree to 12 m; trunk well developed. On edge of Camp Creek.

Ficus opposita Miq. var. *indecora* (A. Cunn.) Corner
Sparsely branched shrub/tree up to 8 m (2-8 m); stems and leaves hispid. Growing in vine thicket and on laterite in eucalypt/palm woodland.

Ficus podocarpifolia Corner
Shrub 1-1.3 m. On sandstone below Mitchell Falls close to water.

Ficus platypoda A. Cunn. var. *platypoda.*
Rock fig to 3.5 m. Growing in crevices and on sandstone.

Ficus racemosa L. var. *racemosa.*
Tree to 5 m; bark dark; leaves shiny; bracts brown. Beside creeks.

Ficus virens Ait.
Tree to 5 m. Above Mertens Falls.

Ficus virens Ait. var. *sublanceolata* (Miq.) Corner
Shrub to 4 m; flowers pale green.

Malaisia scandens (Lour.) Planch.
Climber; milky sap. Near lateritic cliff face.

MYRISTICACEAE

Myristica insipida R.Br.
Tree up to 15 m (8-15 m); bark smooth and grey; upper leaves bright green, lower dull; cinnamon coloured tomentum on fruits. In vine thicket and below Mertens Falls.

MYRSINACEAE

Aegiceras corniculatum (L.) Blanco
Mangrove to 3 m; flowers white, in umbels. On landward edge of mangal.

MYRTACEAE

Calytrix achaeta F. Muell.
Shrub to 1.5 m; flowers white. Growing on sandstone and shallow laterite.

Calytrix brachychaeta F. Muell.
Dense shrub to 2.5 m; flowers white/cream. Growing along creek, around falls and on ripped laterite of re-afforestation area.

Calytrix exstipulata DC.
Small, sparsely branched shrub to 1.3 m; flowers pink. Growing on sandstone.

Eucalyptus apodophylla Blakely & Jacobs
Tree up to 12 m (6-12 m); bark smooth and white to grey in colour and peeling in sheets to either white or light pink-orange. In wet area near swamp.

Eucalyptus ? apodophylla Blakely & Jacobs KFK 6784.
Tree to 12 m; bark smooth and grey/white in colour, lower bark peeling in large strips to grey/orange. Common in seasonally wet area of alluvium over volcanic and lateritic scree.

Eucalyptus bigalerita F. Muell.
Tree to 8 m, broad at base, bark white and smooth. Growing on black silt flat beside Camp Creek.

Eucalyptus brachyandra F. Muell.
Deciduous tree to 3 m. Growing from crevices in sandstone.

Eucalyptus confertiflora F. Muell.
Tree to 12 m (5-12 m); bark white/grey or white/yellow and flaky at base, smooth on upper limbs; fruits small and papery. Growing on laterite outside vine thicket and near Camp Creek.

Eucalyptus ? confertiflora F. Muell. KFK 6646.
Tree to 8 m; bark grey and slightly flaky; fruits small and papery. Near Camp Creek.

Eucalyptus aff. *dichromophloia* F. Muell. KFK 5042.
Bloodwood, on sandstone plateau above Mitchell Falls.

Eucalyptus foelscheana F. Muell.

Tree to 12 m; bark finely flaking and orange in colour, which extends up onto branches; leaves are pendulous and wide at base tapering to a point, medium to dark green in colour. On exposed Carson Volcanics on east side of Camp Creek.

Eucalyptus herbertiana Maiden

Tree to 5 m; bark smooth grey. Growing on sandstone above Mitchell River Falls.

Eucalyptus latifolia F. Muell.

Tree to 8 m; bark white/orange or orange/yellow and finely flaking. Growing on shallow soil over laterite, forming low open woodland; on sandstone near Mertens Falls.

Eucalyptus ? latifolia F. Muell. KFK 5279.

Tree to 8 m; bark orange and flaky; flowers cream.

Eucalyptus miniata A. Cunn. ex Schau.

Tree to 15 m; lower bark rough, upper smooth, flowers orange. Very common on lateritic plateau.

Eucalyptus nesophila Blakely

Bloodwood up to 16 m; bark orange in colour and finely tesselated; leaves are linear with slightly weeping habit; flowers pale cream/white; fruits small. Growing on residual laterite-capped hills over Carson Volcanics.

Eucalyptus papuana F. Muell.

Tree to 14 m; bark rough or smooth; fruits papery. Growing by Camp Creek.

Eucalyptus ? papuana F. Muell. KFK 6767, 6777.

Tree to 9 m; lower bark grey and peeling in large flakes, upper bark smooth and white/grey in colour. Growing on alluvial soil beside Camp Creek.

Eucalyptus perplexa Maiden & Blakely

Tree to 5 m, bark heavy black and ridged. Growing on lateritic rise in Lawley River valley.

Eucalyptus polycarpa F. Muell.

Tree up to 9 m; bark fibrous, slightly tesselated and grey in colour; flowers cream-yellow with 'burnt sugar smell'; buds silvery lepidote, operculum not well developed. Growing on edge of woodland and on laterite.

Eucalyptus tectifica F. Muell.

Tree up to 12 m; bark light grey and finely tesselated; leaves light green. On Carson Volcanics.

Eucalyptus tetrodonta F. Muell.

Dominant tree with *Eucalyptus miniata*. On lateritic plateau.

Eucalyptus sp. KFK 5262, 6702.

Box to 5 m, bark silver grey and fibrous-platy. Growing on Carson Volcanics at Port Warrender.

Eugenia angophoroides F. Muell.

Tall, straight trunked tree to 20 m; base buttressed, bark grey/brown peeling in flaky chunks. Growing beside creek in basalt gorge.

Eugenia eucalyptoides F. Muell.

Much branched shrub to 4 m; flowers cream. Common in creek bed below Surveyors Pool Falls. Tree to 10 m, bark grey and flaky. Growing on grey silt area of Airfield Swamp.

Melaleuca leucadendron (L.) L. *sens. lat.*

Tree up to 14 m (6-14 m); white bark. Common in Airfield Swamp and by Camp Creek.

Melaleuca viridiflora Soland. ex Gaertn.

Tree up to 20 m (3-20 m). Growing on edge of woodland near Airfield Swamp and on sandstone around Mertens and Surveyors Pool Falls.

Osbornia octodonta F. Muell.

Mangrove to 5 m. On edge of sandy beach, Port Warrender.

Tristania ? suaveolens (Gaertn.) Sm. ASG 13135.

Tristania sp. KFK 6764, 6976.
Tree to 12-14 m; bark dark brown to black, fissured. Common along watercourses on Carson Volcanics.

Verticordia cunninghamii Schau.
Shrub to 2.5 m; flowers white. Growing on white sand on sandstone plateau above Mitchell Falls and on laterite near Airfield Swamp.

Xanthostemon eucalyptoides F. Muell.
Tree to 5 m, bark grey, flowers yellow. Confined to the rugged sandstone adjacent to watercourses of Surveyors Pool Falls.

Xanthostemon paradoxus F. Muell.
Spreading tree up to 7 m; flowers yellow. Common around sandstone outcrops of Mertens Falls and Camp Creek.

NYCTAGINACEAE

Boerhavia diffusa L.
Prostrate herb; leaves with reddish tinge; flowers pinkish. Growing on laterite.

Pisonia aculeata L.
Liane; flowers small and inconspicuous. In vine thicket.

NYMPHAEACEAE

Nymphaea gigantea Hook.
Aquatic; flowers blue. Growing in creek on sandstone.

Ondinea purpurea den Hartog
Emergent, tuberous, perennial aquatic, leaves crenulate, flowers purple. In Mitchell River and tributaries.

OLEACEAE

Jasminum parviflorum Dcne.
Straggly scandent shrub; flowers white and strongly perfumed. In understorey of vine thicket.

? Olax sp. KFK 6602.
Shrub. Growing on laterite near costean.

ONAGRACEAE

Jussiaea suffruticosa L.
Flowers yellow. Amongst grasses in alluvial wet valley.

Ludwigia parviflora Roxb.
Herb; flowers yellow. Growing amongst *Typha* and on edge of 'Caravan Creek'.

OPILIACEAE

Opilia amentacea Roxb.
Shrub. Growing in closed grassland.

PAPILIONACEAE

Abrus precatorius L.
Climber growing on dead trees.

Aeschynomene villosa Poir.
Sprawling sub-shrub to 0.5 m. Growing in wet area with *Typha domingensis*.

Atylosia lanceolata W.V. Fitzg.
Flowers yellow. Growing on shoreline.

Atylosia sp. KFK 5202.
Shrub 1.5 m high; flowers yellow. Growing on coast.

Austrodolichos errabundus (Scott) Verdc.
Scrambling legume, flowers pink. Growing on basalt and on lateritic plateau.

Canavalia papuana Merr. & Perry
Scrambling legume, flowers pink. Growing in vine thickets.

Canavalia rosea (Sw.) DC.
Prostrate legume, flowers pink. Growing behind shoreline at Port Warrender.

Clitoria australis Benth.
Scrambling legume. On plateau.

Crotalaria humifusa Grah.
Prostrate pea; flowers yellow. Growing on damp mud.

Crotalaria linifolia L.f.
In fruit only (pods black).

Crotalaria membranacea W.V. Fitzg.
Sparsely branched shrub to 1 m; flowers yellow.

Crotalaria retusa L.
Tree to 5 m; leaves shiny green; flowering finished. Growing on rocky edge of Camp Creek and on eroding edge of lateritic plateau.

Crotalaria trifoliastrum Willd.
Pea, flowers yellow and brown/orange. Growing at edge of 'Caravan Creek'.

Crotalaria verrucosa L.
Shrub; flowers blue. Growing on coast behind mangroves.

Crotalaria sp. KFK 4867, 5072, 5216, 5334, 6622.
Erect shrub to 1.5 m; flowers yellow/orange; calyx lobes hairy. Growing in closed grassland on edge of vine thicket; on sandstone; under *Pandanus.*

Cyclocarpa stellaris Afz. ex J.G. Baker
Ephemeral legume; flowers yellow; pods moniliform. Forms mat on damp, black silt area.

Desmodium filiforme Zoll. et Mor.
Growing on edge of woodland.

Desmodium muelleri Benth.
Ephemeral; ± erect herb.

Desmodium tortuosum (Sw.) DC.
Shrub to 3 m, pods with pale maroon tinge. Growing on disturbed soil at Amax basecamp.

†*Desmodium triquetrum* (L.) DC.
Shrub to 1 m, flowers pink. Growing on black silt beside Camp Creek.

Erythrina vespertilio Benth.
Tree to 4 m with smooth bark. Growing on Carson Volcanics.

Flemingia involucrata Benth.
Spindly perennial herb; flowers light mauve. Growing beside Camp Creek and on gravelly loam.

Galactia tenuiflora (Willd.) Wight & Arn.
Trailing herb. On plateau.

Indigofera linifolia Retz.
Flowers red. Growing on wet area beside Camp Creek.

Indigofera ? pratensis F. Muell. ASG 14470.
Several shrubs in fruit.

Indigofera hirsuta L.
Flowers yellow. Growing on edge of swamp.

Jacksonia argentea C.A. Gardn.
Sterile shrub. Growing on sandstone near Camp Creek.

Mucuna reptans Verdc.
Creeper along ground; flowers mottled yellow and green; pods with golden indumentum. Growing on laterite and basalt.

Psoralea badocana (Blanco) Benth.
Pea 1 m. Growing on moist soil at edge of stream with *Typha.*

Psoralea leucantha F. Muell.
Shrub to 2 m. Flowers blue/purple fading to white at base of each petal; pods black. Common.

Rhynchosia ? minima DC. ASG 14474.
Perennial scrambler.

Rhynchosia sp. KFK 4807, 5018.
Spindly viscid shrub; flowers yellow; standard and keel yellow in open flower, maroon in bud. Growing on sandstone.

Sesbania cannabina Retz.

Stylosanthes humilis H.B.K.
Flowers yellow. Growing on wet areas near camp.

Tephrosia leptoclada Benth.
Herb with ? perennial stock. Flowers maroon/deep pink.

Tephrosia oblongata R.Br. ex Benth.
Erect shrub to 1.5 m, flowers orange. Growing on sandstone north of Surveyors Pool Falls.

Tephrosia phaeosperma F. Muell. ex Benth.
Shrub to 1.3 m. Growing on edge of 'Caravan Creek'.

Tephrosia rosea F. Muell. ex Benth.
Herb to 1.3 m; flowers pink. Growing above Mitchell Falls.

Tephrosia simplicifolia Benth.
Flowers orange; regenerating on burnt area.

Tephrosia sp. KFK 6692.
Flowers blue. Common.

Zornia muriculata Mohlenbrock
Herb; stock perennial (?).

Zornia sp. KFK 6830.
Flowers yellow.

PASSIFLORACEAE

Passiflora foetida L.
Climber. On edge of vine thickets and in riverine forest along Camp Creek.

PITTOSPORACEAE

Citriobatus spinescens (F. Muell.) Druce
Spindly or compact shrub to 2 m; fruits appear compound. Growing near edge of vine thicket.

PLUMBAGINACEAE

Aegialitis annulata R.Br.
Mangrove to 1 m. On landward edge of mangal, Port Warrender.

PODOSTEMACEAE

†*Tristicha trifaria* (Bory ex Willd.) Spreng.
Submerged aquatic in fast flowing water of Mitchell River above Mitchell Falls.

POLYGALACEAE

Polygala chinensis L.
Herb. Growing on edge of swampy closed grassland.

Polygala leptalea DC.
Pink flowered, usually single stemmed herb. Growing on edge of Airfield Swamp.

PORTULACACEAE

Calandrinia uniflora F. Muell.
Herb; stems red; leaves succulent, terete, suffused red; petals magenta. Common on Carson Volcanics.

Calandrinia sp. KFK 4782, 5357, 6606.
Common herb; flowers pink tinged golden on back, centre white; lower leaves red. Growing on laterite.

Portulaca bicolor F. Muell.
Herb with fleshy leaves; flowers pink. Growing in shallow soil over Carson Volcanics.

Portulaca clavigera Geesink
Prostrate fleshy herb; stems red; flowers bright pink; hairs at nodes whitish. Petals, stamens, stigma and anthers magenta; stamens 15, petals ± 8 mm long. Growing in crevices of Carson Volcanics on east side of creek.

Portulaca pilosa L.
Succulent herb; flowers orange. Growing on white sandy beach, Port Warrender.

PROTEACEAE

Banksia dentata L.f.
Tree/shrub to 5 m; flowers yellow. Growing on sandstone, in wet area with *Eucalyptus apodophylla*, and on edge of swamp.

Grevillea agrifolia A. Cunn. ex R.Br.
Multistemmed shrub/tree to 3.5 m; new leaves pale, ferruginous; buds green; flowers cream, odiferous. Perianth fringed with white hairs; stigma, style and ovary are green, gland is white. Growing on sandstone. Common.

Grevillea cunninghamii R.Br.
Shrub to 2 m. Near Camp Creek.

†*Grevillea dryandri* R.Br.
Sprawling shrub growing in sandstone in lightly wooded area at Lawley River valley.

Grevillea latifolia C.A. Gardn.
Shrub to 2 m; flowers red, some white. Growing in white alluvial sand and on sandstone with *Bossiaea bossiaeoides.*

Grevillea heliosperma R.Br.
Shrub to 6 m, flowers reddish, perianth recurved green, pedicel tinged pink. Growing in patch of vine thicket and near Airfield Swamp.

Grevillea mimosoides R.Br.
Sterile shrub to 3 m. Growing on Carson Volcanics.

Grevillea parallela Knight
Tree to 8 m with grey slightly tesselated bark. Growing on escarpment slope in deep soil.

Grevillea pteridifolia Knight
Tree/shrub to 8 m; flowers bright orange. Scattered on deep grey silt which is seasonally inundated. Growing on edge of swamp and on sandstone near creek.

Grevillea pyramidalis A. Cunn. ex R.Br.
Under-shrub to 3 m; flowers cream. Common on plateau.

Grevillea refracta R.Br.
Erect shrub/tree to 5 m; foliage silvery; roughish bark; flowers orange; perianth red with cream limb and style; pollen presenter red. Growing on lateritic plain and on sandstone.

Grevillea aff. *wickhamii* Meisn. Marshall s.n.
Camp Creek.

Hakea arborescens R.Br.
Tree to 4 m. Growing on sandstone below falls and beside creek.

Persoonia falcata R.Br.
Small tree to 3 m, flowers yellow. Growing in margin of vine forest.

RANUNCULACEAE

†*Clematis pickeringii* A. Gray
Climber. In canopy of vine thicket.

RHAMNACEAE

Alphitonia excelsa (Fenzl) Benth.
Tree to 5 m, common along edge of creek.

Emmenospermum cunninghamii Benth.
Tree to 5 m, flowers orange, seeds red. Growing along sandstone rocks by creek.

Zizyphus quadrilocularis F. Muell.
Tree to 7 m; branches spinescent; bark white/grey.

Zizyphus sp. KFK 5265.
Spinescent shrub to 3 m.

RHIZOPHORACEAE

Bruguiera exaristata Ding Hou
Mangrove to 4.5 m. Growing in tidal zone at Port Warrender.

Carallia brachiata (Lour.) Merrill
Tree to 12 m; corky bark; upper surface of leaves glossy, lower dull.

Ceriops tagal C.B. Rob.
Mangrove to 3 m with buttress roots; flowers white. Growing on saline flat.

Rhizophora stylosa Griff.
Mangrove to 3 m; tangled prop roots. Growing on seaward side of *Aegiceras* and *Xylocarpus.*

RUBIACEAE

Borreria aff. *breviflora* (Benth.) Specht. KFK 4811, 4871, 4890.
Ephemeral, flowers blue. On laterite.

Borreria aff. *geeingwa* Specht ASG 14484.
Ephemeral, flowers mauve. In laterite under *Eucalyptus latifolia.*

Borreria sp. KFK 5302, 6605.
Ephemeral, flowers blue. In swampy closed grassland.

Dentella repens Forst.
Prostrate, succulent herb growing in river washed pisolitic gravel in middle of Camp Creek.

Gardenia megasperma F. Muell.
Tree to 4 m. Growing on sandstone below Mertens Falls.

Gardenia ? megasperma F. Muell. ASG 13130.
Growing at Surveyors Pool Falls.

Gardenia pantoni F. Muell.

Gardenia sp. KFK 4797, 5012, 5116, 6746.
Tree/shrub to 3 m; bark orange/red; immature fruits green. Growing on laterite.

†*Knoxia stricta* Gaertn.
Herb; leaves soft; flowers pale blue. Growing on lateritic plateau.

Nauclea coadunatus J.E. Sm.
Shrub/tree to 7 m; leaves dark green. Growing by edge of creek.

Oldenlandia galioides (F. Muell.) F. Muell.
Herb. Growing near kitchen of main camp.

Pavetta granitica Bremerk.
Tree to 4 m, fruits green. Growing in damp seepage area S.W. of Amax Campsite.

Pavetta sp. KFK 7099.
Tree to 4 m, bark dark. Growing in 'Crusher Vine Thicket'.

Randia cochinchinensis (Lour.) Merrill
Shrub/tree to 12 m; upper leaves bright green, lower dull; fruits orange/red. Common in vine thicket.

Timonius timon (Spreng.) Merrill
Tree to 6 m; bark brown; flowers white and falling. Growing alongside creek in alluvial valley.

RUTACEAE

Boronia lanuginosa Endl.
Herb, flowers pink. Growing on sandstone near Mitchell Falls.

Glycosmis pentaphylla (Retz.) DC.
Tree/shrub to 3 m. Understorey in vine thicket.

Micromelum minutum (Forst.f.) Wight & Arn.
Tree/shrub to 4 m; fruits green. Growing in vine thicket.

Murraya paniculata (L.) Jack var. *ovatifoliolata* Engl.
Tree/shrub to 2 m; fruits scarlet, succulent.

Paramignya trimera (Oliv.) Burkill
Shrub; leaves with white galls. Growing in vine thicket.

SANTALACEAE

Exocarpos latifolius R.Br.
Tree/shrub to 2 m. Growing on sandstone.

SAPINDACEAE

†?*Alectryon serratus* Radlk. KFK 5103, 6675.
Under-shrub to 0.5 m; young leaves plum-red. Growing in vine thicket and in vine thicket on edge of plateau.

Atalaya hemiglauca F. Muell. ex Benth.
Shrub to 4 m. Growing on edge of vine thicket.

Atalaya variifolia (F. Muell.) F. Muell. ex Benth.
Tree/shrub to 7 m. Growing in vine thicket.

Atalaya ? variifolia (F. Muell.) F. Muell. ex Benth. KFK 6652.
Tree to 7 m, common in vine thicket.

Distichostemon hispidulus (Endl.) Baill.
Shrub to 0.5 m; stigmas red. Growing on sandstone.

Dodonaea lanceolata F. Muell.
Shrub to 1 m. Growing on basalt near creek.

Dodonaea sp. A.KFK 7079.
Much branched shrub to 1 m. Growing on edge of Camp Creek in exposed basalt.

Dodonaea sp. B.KFK 7037.
Spindly shrub to 1 m, fruits winged. Growing in gorge downstream from Camp Creek.

†*Ganophyllum falcatum* Bl.
Tree to 16 m; bark grey to orange brown; leaves shiny green. Growing on sandstone and in vine thicket.

SAPOTACEAE

Mimusops elengi L.
Tree/shrub 2-5 m; flowers inconspicuous; fruits cinnamon coloured turning bright orange. Growing in remnant patch of vine thicket at Port Warrender.

Planchonella pohlmanniana (F. Muell.) Pierre ex Dubard

Tree to 12 m; fruit large, yellowish/green. Growing on sandstone in vine thicket and at Surveyors Pool Falls.

Pouteria sericea (Ait.) Baehri.

Tree to 12 m; leaves velvety, pubescent. Growing on sandstone in vine thicket and at Surveyors Pool Falls.

SCROPHULARIACEAE

Buchnera linearis R.Br.

Erect herb; flowers mauve. Growing on laterite and on Carson Volcanics.

?*Buchnera linearis* R.Br. KFK 5181.

Herb, flowers mauve. Growing in Carson Volcanics.

†*Centranthera cochinchinensis* (Lour.) Merrill

Herb to 30 cm, flowers maroon, corolla ridged and falling readily. In closed swampy grassland under *Eucalyptus apodophylla.*

Hemiarrhena plantaginea Benth.

?Ephemeral herb, usually rosetted; flowers blue with paler throat. Growing in swampy areas.

Limnophila sp. KFK 4941, 5333, 6782/A.

Emergent aquatic; flowers lilac. Growing in pools 0.5 m deep and forming dense meadow in creek with *Eriocaulon.*

Mimulus sp. KFK 7112.

Decumbent herb; flowers white. Growing in grey silt which is seasonally waterlogged in sandstone.

Mimulus sp. KFK 7029.

Ephemeral herb; flowers violet with white throat. Growing in shallow soil under *Eucalyptus latifolia.*

Mimulus sp. Maconochie 1282.

Erect herb to 10 cm; flowers blue. Growing in black clay soil in fresh water at Surveyors Pool.

Stemodia coerulea Benth.

Flowers blue. Common in patches in woodland.

Stemodia grossa Benth.

Flowers mauve. Growing on sandstone.

Stemodia lythrifolia F. Muell.

Spindly shrub to 1 m; flowers blue/mauve. Growing on sandstone.

Stemodia sp. KFK 6618.

Flowers blue, common in patches of palm/eucalypt woodland.

Striga sp. KFK 7068.

Erect herb, flowers pink. Very common in closed grasslands on plateau.

Striga sp. A.KFK 4776, 4893, 5311, 5336, 6577, 6621.

Erect shrub to 0.5 m; flowers white suffused mauve. Growing on laterite and in closed grassland (*Eriachne pauciflora*).

Striga sp. B.KFK 5331, 6585. J. Lewis 7, 22/A.

Herb; flowers white, slightly fragrant, turning black when dead. Growing on sandstone and in swamp.

Striga sp. C.KFK 5358, 6612.

Flowers pinkish, turning blue with age. Common on lateritic plateau.

SOLANACEAE

Physalis minima L.

Growing around campsite, probably introduced.

Solanum dioicum W.V. Fitzg.

Spindly herb growing on sandstone depression below Mertens Falls.

Solanum vansittartensis C.A. Gardn.
Erect, spindly semi-woody shrub to 2 m; flowers blue. Very common in patches of sand on sandstone.

SONNERATIACEAE

Sonneratia alba J. Sm.
Mangrove to 7 m with thick conical pneumatophores. On seaward side of mangrove belt.

STACKHOUSIACEAE

Stackhousia viminea Sm.
Erect herb; flowers yellow. Growing amongst grass on plateau and on perimeter of swamp.

Stackhousia sp. KFK 6728, 7064.
Erect herb; flowers yellow. Common in wet area.

STERCULIACEAE

Brachychiton diversifolium R.Br.
Tree to 6 m, sterile. Growing on Carson Volcanics and on lateritic hill, grassy area with *Planchonia* and *Terminalia canescens.*

Brachychiton paradoxum Schott
Deciduous tree to 12 m; large leaves; immature fruit with golden tomentum. Growing on edge of vine thicket.

Brachychiton sp. KFK 7056.
Tree to 4 m, bark grey. Growing on sandstone east of Surveyors Pool Falls.

Melochia corchorifolia L.
Woody herb to 1 m; flowers purple-red. Forming small stands on laterite in closed grassland adjacent to airfield.

Melochia umbellata (Mout.) Stapf.
Tree/shrub to 7 m. Common in vine thicket.

Sterculia quadrifida R.Br.
Deciduous tree to 12 m; immature fruit green; mature fruit scarlet, seed black. Bark silvery; flowers green. Growing in vine thickets.

Waltheria indica L.
Flowers yellow.

STYLIDIACEAE

Stylidium alsinoides R.Br.
Decumbent herb, stems and calyx red, flowers pink, throat yellow. In swampy closed grassland under *Eucalyptus apodophylla.*

Stylidium aff. *ceratophorum* O. Schwartz KFK 6716.
Herb, flowers orange pink, back of perianth pale pink, throat appendages yellow. In swampy closed grassland.

Stylidium claytonioides W.V. Fitzg.
Basally tufted herb, stem red, leaves green tinged red, flowers magenta with yellow throat. In swampy closed grassland.

Stylidium fissilobium F. Muell.
Herb, flowers white/mauve. In swampy closed grassland.

Stylidium floodii F. Muell.
Herb, flowers pink. Common in watered lawn near Mining Camp.

Stylidium irriguum W.V. Fitzg.
Herb, flowers white. On edge of Camp Creek.

Stylidium multiscapum O. Schwartz
Herb with basal rosette, flowers deep pink. Common in watered lawn near Mining Camp.

Stylidium rotundifolium R.Br.
Herb, flowers white. Common in swampy closed grassland.

Stylidium schizanthum F. Muell.
Basally rosetted herb, flowers pink. Common in swampy closed grassland.

Stylidium uliginosum Swartz
Basally rosetted herb, flowers pink. Common in swampy closed grassland.

Stylidium sp. KFK 4754.
Basally rosetted herb, with fine, almost terete leaves; flowers white with slight pink tinge. Growing on edge of Airfield Swamp.

THYMELAEACEAE

Pimelea concreta F. Muell.
Herb, leaves light green, flowers white occasionally tinged pink. Growing on laterite in palm woodland.

Pimelea sanguinea F. Muell.
Erect plant; lower stem red, upper stem and leaves green; bracts green; flowers deep red. Growing on laterite with *Hibbertia.*

TILIACEAE

Grewia polygama Roxb.
Spindly shrub. Growing on laterite at edge of woodland.

Grewia sp. KFK 5128, 6619, 6993, 7004, 7097.
Tree to 10 m; fruits green. Growing in vine thicket and in woodland.

Triumfetta aff. *micrantha* F. Muell. KFK 7139.
Much branched shrub to 0.5 m, fruits green. Growing on sandstone S.W. of Mining Camp.

Triumfetta plumigera F. Muell.
Spindly herb to 1 m. Growing on sandstone near creek.

ULMACEAE

Celtis philippensis Blanco
Spiny dense shrub/tree to 15 m; bark grey smooth slightly fissured; developing buttress roots. Growing on coast behind mangroves and in vine thicket.

Trema aspera (Brongn.) Bl.
Shrub to 1.5 m. Growing under *Pandanus* on wet area.

VERBENACEAE

Avicennia marina (Forsk.) Vierh.
Mangrove to 4 m. Growing on seaward and landward side of mangal.

Clerodendrum floribundum R.Br.
Shrub growing on edge of vine thicket.

Clerodendrum aff. *ovalifolium* (A. Juss.) Bakh. KFK 5152.
Tree to 4 m; dry fruits. Growing in vine thicket.

Clerodendrum tomentosum R.Br.
Shrub to 2 m; flowers white. Growing on edge of remnant vine thicket.

Pygmaeopremna subacaulis (F. Muell.) Moldenke
Herb with underground stems; leaves aromatic when crushed. Growing in rocky lateritic soil, in low open forest.

Vitex acuminata R.Br.
Small tree to 5m; fruits black. Growing on sandstone surrounding Surveyors Pool.

Vitex glabrata R.Br.
Tree to 15 m; bark grey, slightly tesselated. Growing against lateritic cliff and on edge of vine thicket.

VIOLACEAE

Hybanthus aurantiacus (F. Muell.) F. Muell.

Herb; flowers yellow. Growing in grey silt (which is seasonally waterlogged) between massive sandstone outcrops.

VITACEAE

Ampelocissus acetosa (F. Muell.) Planch.

Scrambling vine; fruits black.

Cayratia trifolia (L.) Domin

Scrambling vine; common on basaltic scree on edge of plateau, also on sandstone.

DISCUSSION

Vegetation: units, habitats and dynamic status

The vegetation of Mitchell Plateau can be divided into about 20 alliances and sub-alliances that are reasonably well defined floristic entities. These vegetation units may be related, in a predictable manner, to certain characteristic features of their physical environment such as nature of soil and drainage. However, in comparing the vegetation-habitat relationships found by us with those reported elsewhere in northern Australia, it becomes evident that the apparently precise relationships found at Mitchell Plateau do not hold true elsewhere. Instead, on a continental scale there is a very complex relationship between the major vegetation units (alliances and sub-alliances) and the physical environment (climate and soils).

For our study area, the basic pattern of the vegetation mosaic can be summarized as follows: the lateritic plateau supports, where the soil is deep enough and the lateritic cap is sufficiently fractured, an open forest of *Eucalyptus tetrodonta, E. miniata*, and *E. nesophila*. These tree species and their associated understorey species also extend off the plateau on to lateritic scree slopes where a close mosaic with vegetation characteristic of basaltic soils may occur. *Eucalyptus tetrodonta* and *E. miniata* also occur on the surrounding sandstone. On the plateau, where soil is shallow, either shrublands of *Calytrix, Grevillea, Cochlospermum* and *Petalostigma* or low woodlands of *Eucalyptus latifolia* occur. On some parts of the plateau, the *E. latifolia* vegetation appears to follow a remnant drainage system as described by Beard (1976) but frequently no such pattern can be seen. Where active drainage lines occur, *Pandanus* sp. may be found or — in the extreme instance of a large basin-like depression such as Airfield Swamp — permanent wetland vegetation dominated by *Melaleuca leucadendron* occurs. It appears to us that Airfield Swamp is not the "source of Camp Creek" as reported by Beard (1976) although it does overflow into that creek.

Airfield Swamp is of particular interest on Mitchell Plateau where such large bodies of standing fresh water are not common. Six vegetation units have been recognized in the Swamp. Four of these comprise a gradient of water depth, duration of immersion, and soil depth:

(i) *Eriachne pauciflora* alliance (least inundation and very shallow soil),

(ii) *Sorghum* sp. II alliance,

(iii) *Paspalum orbiculare* association of the *Melaleuca leucadendron* alliance, and
(iv) *Eleocharis brassii* association of the *M. leucadendron* alliance.

Within the *Melaleuca* stand where the tree canopy is dense, the understorey is open and the *Eriocaulon setaceum* association occurs. The sixth unit, Closed-Herbland Formation (with a rich assemblage of herbs and graminoides) has been insufficiently studied but occurs primarily on the north side of the swamp where the gradient from deep water to dry land is steep and the *Eriachne pauciflora* alliance is absent. This vegetation unit may be closely related to similar herblands on valley floors found beneath the *Eucalyptus apodophylla* alliance where drainage is impeded and soils are heavy, but detailed studies are needed.

The lateritic plateau was formed on the basalts of Carson Volcanics. A broad fringe of basalt rock and derived basaltic soils surrounds the plateau and supports its own characteristic vegetation. Where the basaltic soils are deep and well-drained, a woodland of *Eucalyptus tectifica* dominates. When further information becomes available, this woodland should be divisible into associations on the basis of understorey species. We observe that rudimentary drainage lines support a suite of species different from those on the surrounding slopes. On the basaltic slopes leading up to the lateritic plateau, the often fragmentary occurrences of the *Eucalyptus confertiflora-E. polycarpa* alliance can be found. *Eucalyptus foelscheana* also occurs along the lower edges of scree slopes below the plateau.

Very gently sloping valley floors, that have accumulated heavy soils, support a *Eucalyptus apodophylla* alliance with its several as yet undescribed herb associations. Drainage in such areas is only moderate but standing water does not appear to be common, and run-off may be rapid after heavy rain. One of the *Pandanus* alliances also occurs in this habitat. Where there is standing water, the *Melaleuca leucadendron* alliance is found. The free-standing water may contain *Philydrum lanuginosum* and *Eriocaulon* spp. The margins of areas supporting the *Eucalyptus apodophylla* alliance frequently support *Themeda australis, Grevillea pteridifolia, Banksia dentata, Planchonia careya* and, more rarely, *Cycas lane-poolei.*

There are areas within *Eucalyptus tectifica* woodland where sheets of basaltic rock are exposed at the surface. Here, in small cracks and depressions, a herbland, the *Portulaca clavigera-Calandrinia uniflora* alliance or the *Eriachne pauciflora* alliance, occurs. Where a thin layer of soil is present, low open woodland of the *Terminalia canescens* alliance with *Gardenia* sp., *Petalostigma pubescens* and *Cochlospermum fraseri* grows.

Areas of alluvial-colluvial soil adjacent to major streams support a mixture of *Terminalia fitzgeraldii* alliance, *Eucalyptus papuana*, and *Pandanus* spp. alliances. The latter two tend to occur nearer streams than the first. *Eucalyptus papuana* can also be found in the wet zone below some patches of vine thicket. Fringing the major streams is a riverine zone with several tree species of medium height that are related to those of the vine thickets. Also occurring here are stands of *Pandanus aquaticus* and *Pandanus spiralis*, a dense, tall growth of grasses and some herbs, the species of which vary with degree of inundation and soil type.

Two *Sorghum* species alliances have been recognised. One forms a conspicuous zone peripheral to *Melaleuca leucadendron* at Airfield Swamp, and the other is most frequently found as a zone peripheral to many vine thickets. *Sorghum* is also an early invader of disturbed ground such as grader spoil beside tracks. The current systematic difficulties of the genus *Sorghum* make it impossible to understand clearly the role of *Sorghum* in this area, and more work is needed to clarify the situation.

Surrounding the area of volcanic rocks is a wide expanse of sandstone country. We examined a few parts of it only briefly during our study. Most common is the *Eucalyptus herbertiana* alliance on areas of rugged sandstone. Because of their relative floristic richness, the areas of permanent or semi-permanent water along water courses are of interest. Here the biogeographically isolated patches of Gondwanan plants, related to those of the vine thickets, are found. The steep-walled gorges carved by the rivers and streams create either locally humid environments in which ferns grow, or else harsh conditions where plants such as *Hoya australis* and *Calytrix exstipulata* grow from small pockets of soil. The alluvial flats appear to offer yet another habitat. Much detailed work could be done on the sandstone country so as to permit a clear understanding of the relationships between species, vegetation and environment.

The north-eastern edge of Mitchell Plateau overlooks Admiralty Gulf, a large embayment along the southern edge of the Timor Sea. The sheltered strand line of this tropical gulf supports, in several places, well-developed mangrove vegetation. The mangrove sub-alliance described in this report is seen on aerial photographs to be segregated into well defined zones parallel to the coast and along major streams. From our examination of one area of mangroves to the south-west of Walsh Point, we found the zonation **(Table 25)** that appears to be similar to that around the whole Gulf.

The mangrove vegetation occurs as discontinuous, elongated patches that are characterized by a prominent ridge formed by the massive *Rhizophora* trees (see **Table 8** *Rhizophora stylosa* sub-alliance). At the lateral edges of the patches, the *Ceriops-Avicennia* sub-alliance extends around the edges of the more seaward zones. It is in these lateral areas that several uncommon species associated with either mangrove or non-mangrove, tropical strand lines were found, e.g. *Bruguiera, Thespesia, Canavalia, Sporobolus, Osbornia.*

Our relatively short periods on the plateau and in its vicinity allow us to make the following observations and tentative comments relating to the dynamic status of some of the vegetation units. The vegetation of Mitchell Plateau may, to a large degree, be representative of climatic climax plant communities. On the plateau itself, specimens of the abundant understorey palm, *Livistona eastonii* may be about 280 years old (Hnatiuk, 1977), perhaps suggesting that the overstorey of *Eucalyptus tetrodonta, E. miniata* and *E. nesophila* is at least as old. Since no other tree species appear able to replace these, and since these species regenerate well in disturbed sites, the assessment of this vegetation as the climax appears reasonable. Furthermore, because the *E. tetrodonta-E. miniata-E. nesophila* alliance appears to be in a stable relationship with most of the adjacent alliances, it is assumed that these also are climax communities.

The one major exception to this presumed climax status is the vine thicket alliance. Several circumstantial observations suggest that the vine thickets are remnants of a vegetation that was once more widespread than now, characteristic of a climate more humid than at present and now represented only by fragmented relicts. The evidence for this is:

(i) the occurrence of isolated, fire-scarred thicket trees in a peripheral zone outside the thickets,
(ii) the reproduction of thicket species in the peripheral zone, and
(iii) the sharply delineated, fire-produced boundaries of the thickets.

There is no palaeobotanical information available to date on the north-west of Australia; however for north-eastern Australia, Kershaw (1971) has shown that about 3,000-2,000 B.P., rainforest rather than eucalypt-dominated sclerophyll vegetation was the principal vegetation type, probably because of higher temperature and precipitation. Since 2,000 B.P. the rainforests near Quincan Crater, Qld., have reflected a drier (*sic*) character and have been replaced in part by sclerophyll vegetation. If the Kimberley had a similar climatic history then our hypothesis of the current remnant status of vine thickets in our area would be plausible.

The vine thickets of our study area all seem to occur in places with particularly favourable access to ground water: at the base of the massive lateritic capping of the plateau and at the base of lava flows overlooking Port Warrender. Also, species with biogeographic affinities similar to those of vine thicket species occur as a narrow riverine fringe in several places. Not all places with presumed favourable water relations support vine thickets. In some areas, a dense growth of eucalypt-dominated vegetation grows in such locations and its relative luxuriance indicates the favourable nature of the site.

The role of fire in relation to vine thickets has been explored by Beard (1976) who notes that vine thickets are often protected from fire because they commonly occur below barren rocky cliffs and amongst large-boulder screes. There are several stands of vine thicket, however, which do not owe their existence to special fire protection, e.g. those along parts of the western escarpment above the headwaters of the Surveyors Pool stream. These patches are surrounded by 'eucalypt-grassland', are often free of boulder scree and are isolated from the cliff face, which may be very low in any case. Frequently they are partially or entirely surrounded by a dense stand of *Sorghum* sp. 2-2.5 m tall, a feature shared with many other vine thickets and one requiring a special mention.

These *Sorghum* belts presumably indicate the favourable water supply associated with vine thickets' habitat. When ignited in the dry season, they are completely consumed in a conflagration of considerably greater intensity than would occur with the usual 'eucalypt-grassland' understorey vegetation. What effect do such fires have on vine thickets? Some tree species of the vine thicket can be found growing in the *Sorghum* zone (e.g. *Albizia lebbek, Dolichandrone heterophylla*) indicating that the thickets were once larger than at present and that some vine thicket trees can survive fire. It appears that there may have been more frequent fires on the plateau during the past decade than before (Hnatiuk, 1977). It is our hypothesis that the extensive

Sorghum zone represents a response to frequent fires which have encroached on the vine thicket. Where thicket species once grew, *Sorghum* now grows, as it can regenerate under an annual fire regime and is favoured by ample soil water. The frequent and intense fires prohibit regeneration of either vine thicket species or eucalypt woodland species **(Figs 8 & 9)**. However in stands of *Sorghum* unburned for one season, *Sorghum* regrowth appeared retarded by the accumulation of dead grass which did not entirely decompose during the wet season. Seedlings of vine thicket species were present, indicating the potential for regeneration of the thicket community on these sites. Obviously more study is needed.

We have a little information on the response of the vegetation to disturbance, but more detailed observations are needed. The frequent and now virtually annual burning appears to be resulting in a depletion of understorey shrubs. Species such as *Grevillea pyramidalis* are frequently found only as regrowth less than 30-50 cm tall, from underground parts. Many grasses and herbs appear able to regenerate under the annual burning. It is not known whether any species are unable to regenerate under an annual fire regime.

Disturbance of the surface soil by graders and bulldozers results in the obvious and immediate destruction of vegetation in their path. However, if left alone afterwards the flora at such sites appears likely to regenerate to something approximating the original vegetation when the area of disturbance is small or at least narrow within undisturbed areas. Extensive and intensive disturbance such as that caused by mining has a more profound effect on the vegetation, as has occurred at the trial mine site south-west of Mining Camp. Regeneration of some native species (e.g. *Livistona eastonii, Eucalyptus tetrodonta, Gomphrena canescens*) occurs but extensive areas of barren ground persist, perhaps because sheet erosion prevents successful seed implantation. The effect on regrowth of the level of available nutrients in the pallid, clay layer is not known but needs study.

Disturbance of large areas of vegetation, for example by cyclones, appears to initiate the development of a series of regeneration communities that are poorly known at present. Evidence for their existence comes, for example, from an area south-east of the airfield, where a large area of regenerating eucalypts is to be found. A tall stand of *Grevillea parallela* also occurs here and may be seral.

The flora and its biogeographic affinities

Only a small amount of information is available about the non-vascular flora. A number of bryophyte collections were made but as these have not yet been identified they are not included in the list. The lichen list was compiled by N. Sammy (present address: Dampier Salt, Dampier, W.A. 6713) who comments: "The range of material is certainly nowhere as rich as that recorded from the Prince Regent River Reserve (Miles & Burbidge, 1975) but with a predominance of crustose species . . . There appears to be a lack of true tropical foliose *Parmelia* species like *P. cristifera* and *P. tinctorium*, which implies a drier tropical climate." The scarcity of lichens in open vegetation of the dry parts of Arnhem Land, N.T. and Weipu, Qld. has been noted by Specht *et al.* (1977).

Our list records a number of plants previously unpublished for Western Australia. They include 32 taxa but the undetermined material may contain others. Nine of the new records are genera previously not known to occur in the State, viz. *Fissistigma* (Annonaceae), *Garuga* (Burseraceae), *Hyptis* (Lamiaceae), *Knoxia* (Rubiaceae), *Lagerstroemia* (Lythraceae), *Pachygone* (Menispermaceae), *Tristicha* (Podostemaceae), *Alectryon* and *Ganophyllum* (Sapindaceae). The first record of the family Podostemaceae for Western Australia comes from the collection of the aquatic *Tristicha* from the Mitchell River above the falls. *Hyptis suaveolens* was only recorded from around Mining Camp. This suggests that it has been introduced to the plateau. It is an aggressive species and has the potential to become as troublesome here as it is in the Northern Territory. The other 5 genera were all collected from vine thickets. As most of our collecting was done in the dry season, many new species and records could be expected to result from additional collecting during the wet. Many of the collections made by us from vine thickets involved sterile material and in some cases new records are based on fruiting material e.g. *Clematis pickeringii* and *Dioscorea transversa.*

The plant collections made from Mitchell Plateau and the surrounding area include many new records for the State, which support the conclusions of earlier workers (Mueller, 1867; Diels, 1906; Gardner, 1923, 1944, 1959; Burbidge, 1960) that although the Kimberley flora contains a major Australasian (= Autochthonian) element (Specht, 1979). It also includes a number of species whose affinities lie with the Gondwanan (= Indo-Malesian) flora (Specht *ibid.*) (Mueller, 1867; Diels, 1906; Gardner, 1923, 1944 and 1958; Burbidge, 1960).

The Gondwanan element is best developed within the vine thickets, fringing riverine forests and mangals; the mangrove species probably reached Australia by water dispersal. The Gondwanan element within the vine thickets and riverine forests is probably related to the availability of soil water in the dry season. Vine thickets are found primarily in places that have specially favourable access to ground water, such as seepage zones just below the edge of the lateritic plateau and below the basaltic benches away from the plateau. The riverine and vine thicket communities are less susceptible to fire. Unlike the communities dominated by the more open-canopied Australasian element their closed canopies inhibit the development of a grass understorey.

The bright green canopies of the vine thickets and riverine forests contrast with the grey-green leaves of species forming the Australasian element. Deciduous emergent trees such as *Sterculia, Terminalia* and *Bombax* occur more frequently in the vine thickets than in the riverine forests. However, the riverine forests possess more broad-leaved evergreen species such as *Calophyllum sil, Ficus racemosa, Nauclea coadunatus* and *Timonius timon* than do the vine thickets. Deciduous species are not restricted to the vine thickets and occur within communities dominated by species representative of the Australasian element, e.g. *Eucalyptus brachyandra, E. confertiflora*, and *Brachychiton* species which grow on the exposed sandstone plateau. The Baobab (*Adansonia gregorii*) is found in our study area only near the coast on basaltic soils at Port Warrender, whereas the Kapok Bush (*Cochlospermum fraseri*) is common on the sandstone and on the plateau where the underlying basalt is exposed; these species, which are both deciduous, form biogeographical links with the floras of Africa and America. The local distribution of *Adansonia* is interesting in that it appears restricted to special sites formerly used by Aborigines. **(Fig. 28)**.

Arborescent vine thickets and riverine species with Gondwanan affinities include *Albizia lebbek, Carallia brachiata, Melaleuca leucadendron* and *Glycosmis pentaphylla.* Certain species of *Ficus* are valued by E.J.H. Corner (pers. comm. 1978) as indicators of plant geographical affinity in the Indo-Australian region. *Ficus racemosa* is a river bank species the distribution of which indicates floristically related rivers. Corner sees the flora of the river systems of Western Australia, Queensland and Northern Territory as indicating a geographical link with Timor, the South-East Celebes, Meranki, Western Malaysia and Peninsular India. He speculates that *Ficus racemosa* (which does not occur in Madagascar) did not originate in Gondwanaland and then spread into Asia, but that it entered Australia across some link with South East Asia when Australia was drifting north and before New Guinea had appeared.

Genera which characterize the Australasian element are best developed on the arid lateritic plateau and surrounding sandstone. Here, the xeromorphic *Eucalyptus* species form the greater part of the arborescent growth with other woody species such as *Acacia, Grevillea* and *Hakea* being common. However a number of arborescent genera with Gondwanan affinities do occur within these communities, e.g. *Cycas, Owenia, Lysiphyllum, Planchonella, Terminalia* and *Xanthostemon.*

Links with the Australasian flora of south-western Australia are seen in such genera as *Banksia, Byblis, Thysanotus* and *Stylidium* which occur in swampy closed-grasslands where the soil is perennially moist or seasonally inundated. Other predominantly south-western genera include *Boronia, Bossiaea, Calytrix, Hibbertia, Jacksonia, Lomandra, Persoonia* and *Verticordia* all of which occur in the *Eucalyptus* understorey.

ACKNOWLEDGEMENTS

We thank Amax Exploration (Australia) Inc. for permission to use their facilities at Mitchell Plateau. In particular we are grateful to Ralph Day of Amax, who gave us every possible assistance. At the plateau, Amax field staff Morrie Marshall, Duncan Monroe, Symon Platt and Joe Smith provided willing assistance and enabled us to cover much of the plateau.

Technical assistance in the field was provided by Bernie Haberley in 1976 and Karl Pirkopf in 1978 and 1979. We thank them for their willing co-operation and companionship in the field.

Because of the uncertain status of many of the plant groups collected from the plateau we asked for, and received, expert assistance from many of our colleagues both in Australia and overseas. They include E.J.H. Corner, Cambridge University (*Ficus*); P. Taylor, Royal Botanic Gardens, Kew (*Utricularia*); B. Verdcourt, Royal Botanic Gardens, Kew (Leguminosae); B. Stone, University of Malaysia (*Pandanus*); N. Byrnes, Queensland Herbarium, (*Terminalia*), B. Hyland, Division Forest Research, Atherton, Queensland (rain forest species); R.N. Hilton, University of Western Australia (fungi); N. Sammy, Dampier Salt Ltd (lichens); H. Aston, National Herbarium, Victoria (aquatics); M. Lazarides, Herbarium Australiense (Poaceae); S.G.M. Carr, School Developmental Biology, A.N.U. (*Eucalyptus*);

M.I.H. Brooker, Division of Forest Research, C.S.I.R.O. (*Eucalyptus*); D. McGillivray, National Herbarium of N.S.W. (*Grevillea*) and Hj. Eichler, Herbarium Australiense (*Clematis*). We would also thank our colleagues A. George, N. Lander, N. Marchant, B. Maslin, Gillian Perry and P. Wilson for their assistance with various specialist groups.

J. Green and A. George read the manuscript and made valuable comments. Professor Corner made valuable comments on aspects of biogeography and allowed us to cite his unpublished work. Professor R.L. Specht's suggestions in reviewing this paper were appreciated. Wendy Lee-Frampton, Cheryl Lynch and Suzanne Curry provided expert technical assistance in processing the collections and despatching loans.

We thank J. Beard, J. Lewis, and members of the Western Australian Museum survey teams who made plant collections from the plateau.

The drafts and the final manuscript have been patiently and expertly typed by Vicki Hamley to whom we express our thanks.

REFERENCES

ALLEN, A.D. (1971)—*Montague Sound, Western Australia, 1:250,000 geological series explanatory notes.* Canberra: Bureau of Mineral Resources.

APLIN, T.E.H. (1977)—The Vegetation of Western Australia. *Off. Yb. West. Aust. (n.s.) no. 16.*

BEARD, J.S. (1976)—The monsoon forests of the Admiralty Gulf, Western Australia. *Vegetatio* **31:** 177-192.

BLAKE, S.T. (1968)—A revision of *Melaleuca leucadendron* and its allies (Myrtaceae). *Contr. Qd. Herb.* **1:** 1-114.

BURBIDGE, N.T. (1960)—The phytogeography of the Australian region. *Aust. J. Bot.* **8:** 75-211.

BURBIDGE, A.A. & McKENZIE, N.L. *eds.* (1978)—The islands of the North West Kimberley, Western Australia. *Wildl. Res. Bull. West. Aust.* no. 7: 1-47.

DIELS, L. (1906)—Die Planzenwelt von West-Australien südlich des Wendekreises. In: *Die Vegetation der Erde* **7:** 413; ed. A. Engler and O. Drude. Leipzig; W. Engelmann.

GARDNER, C.A. (1923)—Botanical notes, Kimberley Division of Western Australia. Perth. *Bull. For. Dept. West. Aust.* no. 32.

GARDNER, C.A. (1944)—The vegetation of Western Australia. *J. R. Soc. West. Aust.* **28:** 11-87.

GARDNER, C.A. (1959)—The vegetation of Western Australia. Pp. 274-282 in *Biogeography and ecology in Australia*; ed. A. Keast *et al.* The Hague: W. Junk.

GELLATLY, D.C. & SOUFOULIS, J. (1969)—*Drysdale and Londonderry, Western Australia, 1:250,000 geological series explanatory notes.* Canberra: Bureau of Mineral Resources.

GEORGE, A.S. (1978)—Notes on the vegetation and flora of the Cape Londonderry peninsula, North Kimberley, Western Australia. *West. Aust. Herb. Res. Notes* **1:** 1-15.

HALL, N. *et. al.* (1970-77)—*Forest Tree Series* Nos. 1-204. Dept. Natl. Develop. For. Timber Bur. Canberra.

HNATIUK, R.J. (1977)—Population structure of *Livistona eastonii* Gardn., Mitchell Plateau, Western Australia. *Aust. J. Ecol.* **2:** 461-466.

KABAY, E.D. & BURBIDGE, A.A. eds. (1977)—A biological survey of the Drysdale River National Park, North Kimberley, Western Australia. *Wildl. Res. Bull. West. Aust.* no. 6: 1-133.

KERSHAW, A.P. (1971)—A pollen diagram from Quincan Crater, north-east Queensland, Australia. *New Phytol.* **70:** 669-681.

MILES, J.M. & BURBIDGE, A.A. (1975)—A biological survey of the Prince Regent River Reserve, North-West Kimberley, Western Australia. *Wildl. Res. Bull. West. Aust.* no. 3: 1-114.

MUELLER, F. von (1867)—Australian vegetation: pp. 251-268 in *Intercolonial Exhibition Essays 1866-67* no. 5.

PICHI SERMOLLI, R.E.G. (1977)—Tentamen Pteridophytorum genera in taxonomicum ordinen redigendi. *Webbia* **31:** 313-512.

SPECHT, R.L. (1979)—Evolution of the Australian flora: some generalizations. In: *Ecology & biogeography in Australia*; ed. A. Keast *et al.* The Hague: W. Junk.

SPECHT, R.L., ROE, E.M. & BOUGHTON, V.H. eds. (1974)—Conservation of major plant communities in Australia and Papua New Guinea. *Aust. J. Bot. (Suppl.)* no. 7.

SPECHT, R.L., SALT, R.B. & REYNOLDS, S.T. (1977)—Vegetation in the vicinity of Weipa, North Queensland. *Proc. R. Soc. Qd.* **88:** 17-38.

SPECK, N.H. (1960)—Vegetation of the North Kimberley area, W.A. *Land Res. Ser. C.S.I.R.O. Aust* no. 4: 41-63.

STEENIS, C.G.G.J. van (1951)—Flora Malesiana, present and prospects. *Taxon* **1:** 21-24.

Received 6 February 1979 Accepted 23 October 1979 Published 28 February 1981

PART 3

A SURVEY OF RIVERSIDE MANGROVE VEGETATION FRINGING TIDAL RIVER SYSTEMS OF KIMBERLEY, WESTERN AUSTRALIA

A.G. WELLS*

ABSTRACT

Distribution, assessment of cover abundance and pertinent ecological observations are given for mangrove species fringing five tidal river systems in Kimberley, Western Australia.

Fourteen species of mangroves were observed, one of which (*Xylocarpus granatum*) has not previously been recorded from Western Australia.

INTRODUCTION

Limited information is available on the mangrove flora of the Kimberley region in Western Australia. Brief reports have been provided by Gardner (1923), Sauer (1965), Beard (1967) and Perry (1970). Miles *et al.* (1975), Thom *et al.* (1975), Messel *et al.* (1977) and Semeniuk *et al.* (1978) have provided more detailed accounts of tidal wetland plant distibution of the area.

The region described here includes five tidal river systems along the north-west Kimberley coast of Western Australia (**Fig. 1**). All systems are located within deep structural embayments. Floodplain development is minimal and tides in the area are semi-diurnal with amplitudes of up to 10 m.

The 14 species recognized as mangroves in the study area during this survey are listed below and follow the nomenclature of Jones (1971).

Aegialitis annulata R. Br.
Aegiceras corniculatum (L.) Blanco.
Avicennia marina (Forsk.) Vierh.
Bruguiera exaristata Ding Hou.
Bruguiera parviflora (Roxb.) Wright and Arn ex Griff.
Camptostemon schultzii Masters.
Ceriops tagal (Perr.) C B Robinson var. *australis* C T White.
Excoecaria agallocha L.
Lumnitzera racemosa Willd.
Osbornia octodonta F. Muell.
Rhizophora stylosa Griff.

---

*Department of Environmental Physics, University of Sydney, New South Wales. 2006

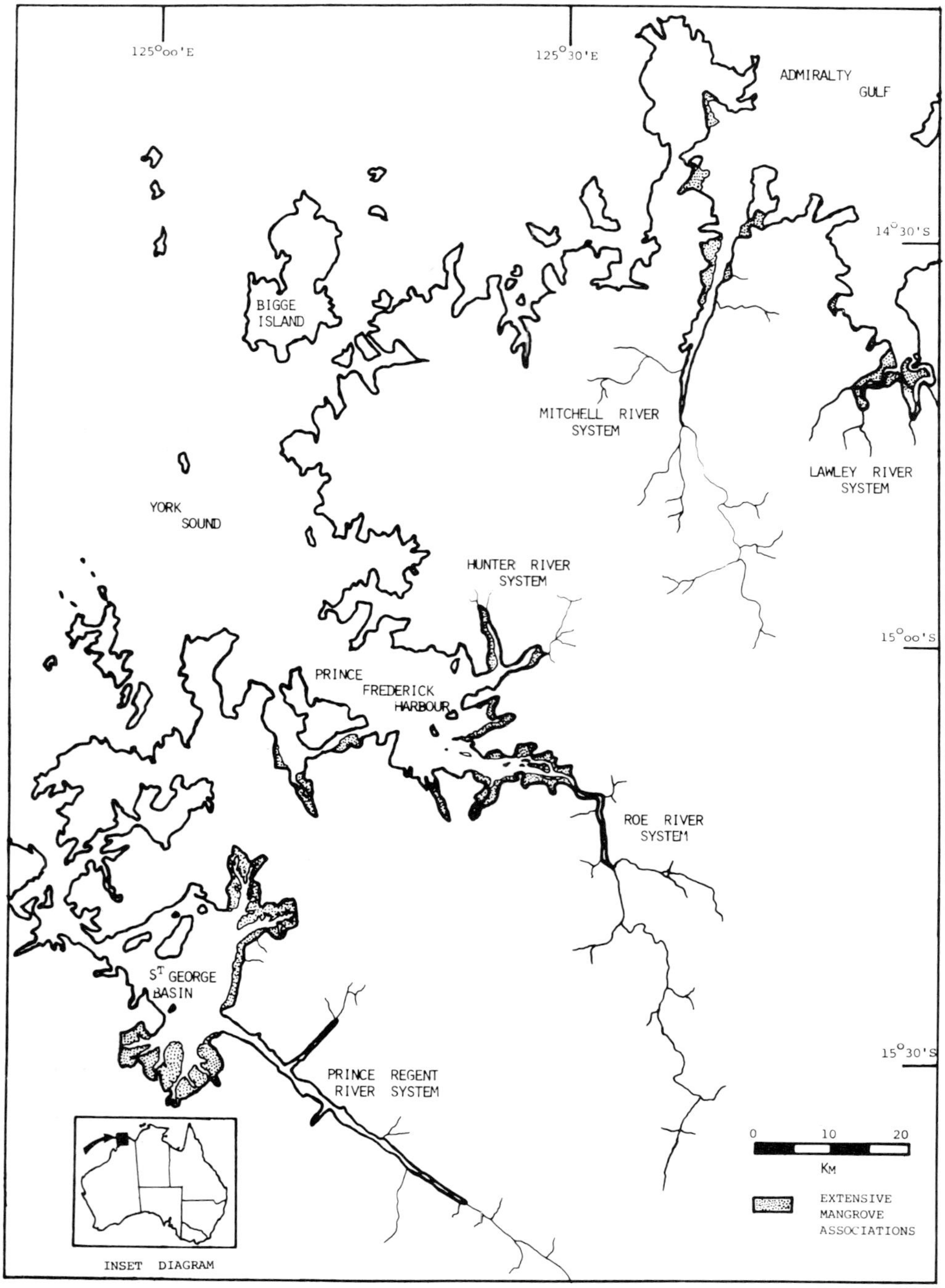

Fig. 1: Study area in the Kimberley District, Western Australia.

Sonneratia alba J E Smith
Xylocarpus granatum Koen.
Xylocarpus moluccensis (Lamk.) Roem. Syn X. *australasicus*? Ridl.

The purpose of this paper is to outline the distribution and abundance of mangrove species occurring as fringing riverside vegetation and to discuss pertinent ecological observations.

METHODS

The study was carried out during July 1977 from the University of Sydney's 21 m Research Vessel *The Harry Messel* while it was involved in surveys to determine the status of the salt-water crocodile in the north-west Kimberley (Messel *et al.*, 1977). Surveys were made from a 5 m boat. Locations were marked on maps prepared from aerial photographs (accurate to ± 0.1 km).

Quadrats 100 m long (along each bank) and 20 m deep (back into the fringing riverside vegetation) were determined at 2.5 km intervals on each river. In areas where rocks abut the banks and the depth of mangrove swamp extends less than 20 m back from the bank, the quadrat is still 100 m long with variable width. The scale of cover abundance used is modified from Braun-Blanquet (1932). It consists of:

1. Cover up to 5%.
3. Cover between 25-50% of area.
4. Cover between 50-75% of area.
5. Cover between 75-100% of area.

High and low tide salinity profiles were taken at 2.5 km intervals along the rivers. Measurements were made at the surface, midwater and bottom using an Autolab (Model 602) salinity/temperature meter.

Lawley River System

The Lawley River, with two major creeks, Mindjau Creek and Rail Creek, **(Fig. 2)**, enters the southern portion of Port Warrender. Tidal range is approximately 7 m. Maximum distance upstream of tidal influence is 10-20 km in all three arms.

Basalt cliffs rise towards the Mitchell Plateau and delimit the catchment areas; 41 km^2 and 149 km^2 of Mindjau and Rail Creeks respectively. The catchment area of the Lawley River—573 km^2—lies for the greater part within Precambrian sandstones.

High and low tide salinity profiles were determined for the 3 tidal arms **(Fig. 5)**. These indicate that even into July (mid dry season), some fresh water is still entering the tidal portion of the river from drainage areas upstream. Mindjau Creek has salinity values typical of seawater at both high and low tides because of its small catchment area.

The Lawley River and Rail Creek show slight decreases in salinity as distance upstream increases, indicating that some fresh water is still entering from the catchment area. Complete vertical mixing of salinity profiles was evident only in Mindjau Creek. Rivers in Arnhem Land, N.T., which have similar salinity profiles in the mid dry season, become hypersaline by the end of the dry season, when fresh water ceases flowing in from the catchment areas. It is considered that all 3 tidal arms of the Lawley River System could become hypersaline by the end of the dry season.

There is approximately 47 km^2 of mangal within the Lawley River-Port Warrender system. The extent of the mangal is:

Lawley River	10.5 km^2	Mindjau Creek	12.0 km^2 (CK A)
Port Warrender	11.5 km^2	Rail Creek	13.0 km^2 (CK B)

Along the Lawley, the dominant riverside mangrove community is a low closed forest (canopy height ≤ 8 m) of *Avicennia marina, Camptostemon schultzii* and *Rhizophora stylosa.* A shrub understorey (canopy height ≤ 2 m) of *Aegiceras corniculatum* occurs along the banks while the pioneer species *Sonneratia alba* (canopy height ≤ 8 m) occurs along convex portions of river meanders and also colonizes accreting mudbanks to seaward.

Distributions and assessments of cover abundance for mangrove species along the tidal arms of the river system are shown in **Fig. 4**.

Around the river mouth, *Sonneratia alba* forms an association with *Avicennia marina* and *Camptostemon schultzii* on extensive gently shelving mudbanks (**Plate 1**). During high tides, these pioneer mangrove associations are inundated nearly to the top of their canopy.

Dominant mangrove vegetation along Mindjau and Rail Creeks was also low closed forests of *Avicennia marina, Camptostemon schultzii* and *Rhizophora stylosa* (canopy height ≤ 8 m), a shrub understorey of *Aegiceras corniculatum* (canopy height ≤ 2 m) being present in both arms. In Rail Creek, *Aegiceras corniculatum* forms an association with *Aegialitis annulata* (canopy height ≤ 0.5 m). Shrubs of *Osbornia octodonta, Aegialitis annulata, Lumnitzera racemosa* and stunted *Avicennia marina* (canopy height ≤ 2.5 m) occupy peripheral positions around many of the saline mudflats.

An association of *Ceriops tagal* var. *australis, Excoecaria agallocha* and *Lumnitzera racemosa* (canopy height ≤ 4.5 m) occur as the landward extension of mangroves bordering the saline mudflats that are colonized in some areas by various samphires—*Arthrocnemum* sp, *Suaeda australis* and *Sporobolus virginicus* being dominant. Most of the saline mudflats are devoid of vegetative cover. As shown in **Fig. 4**, *Xylocarpus moluccensis* was infrequently observed. *Bruguiera* spp. were also sparsely represented in the mangrove forests.

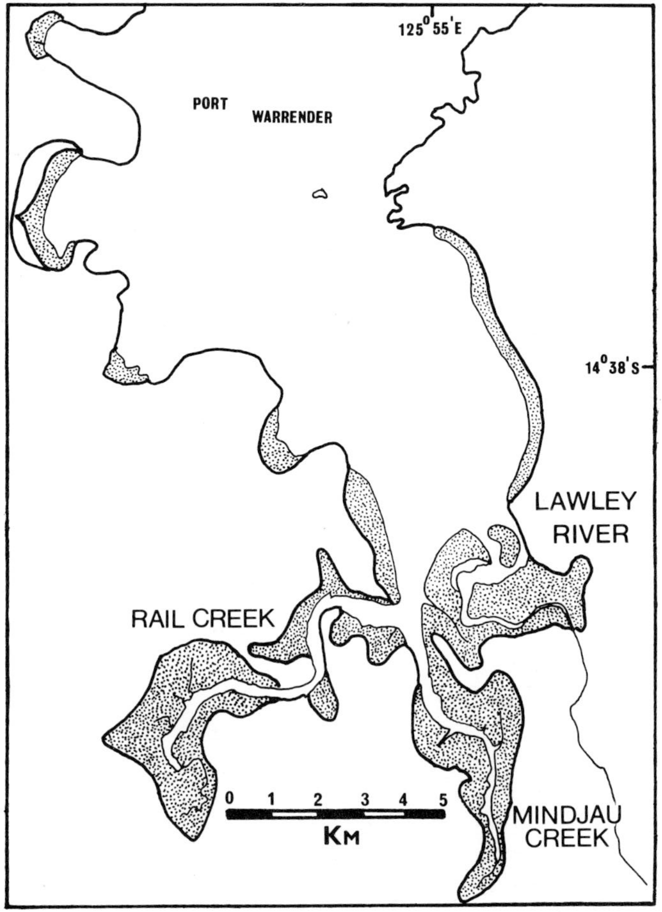

Fig. 2: Port Warrender-Lawley River system showing distribution of mangrove swamps.

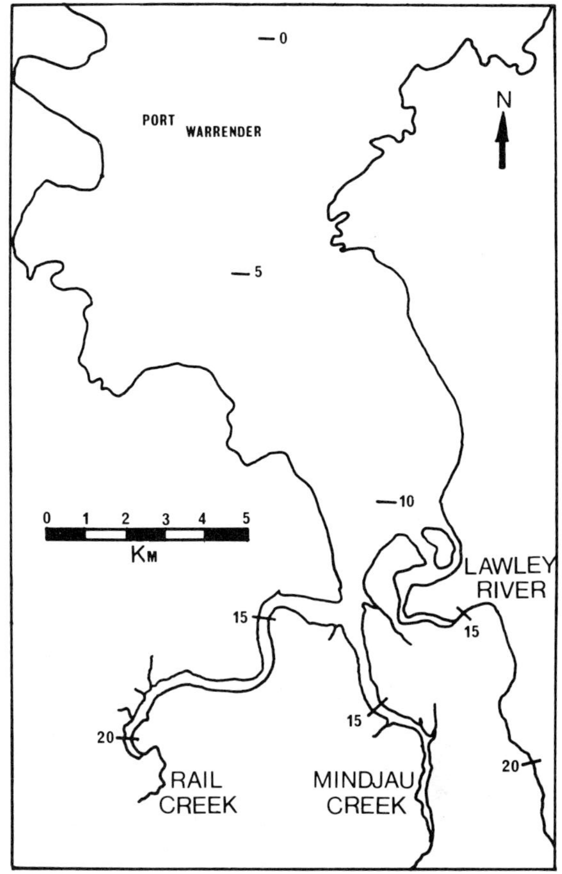

Fig. 3: Port Warrender-Lawley River system showing distances upstream (in km) for location of survey sites.

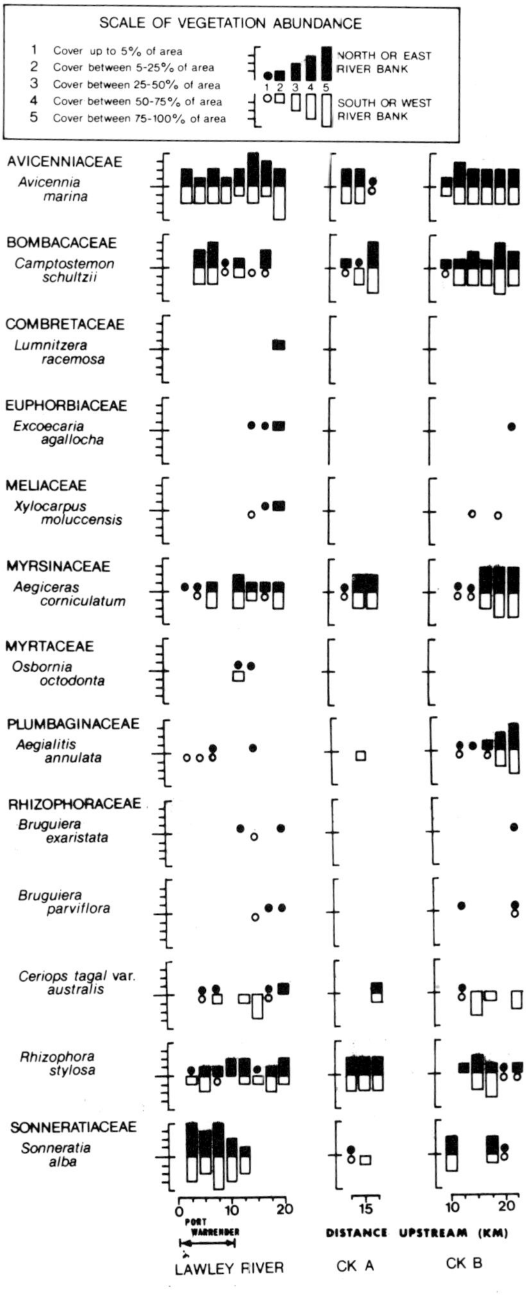

Fig. 4: Distribution and assessment of cover abundance for mangrove species within 100 m (length) x 20 m (depth) quadrats of fringing riverside vegetation along each river bank at 2.5 km intervals.

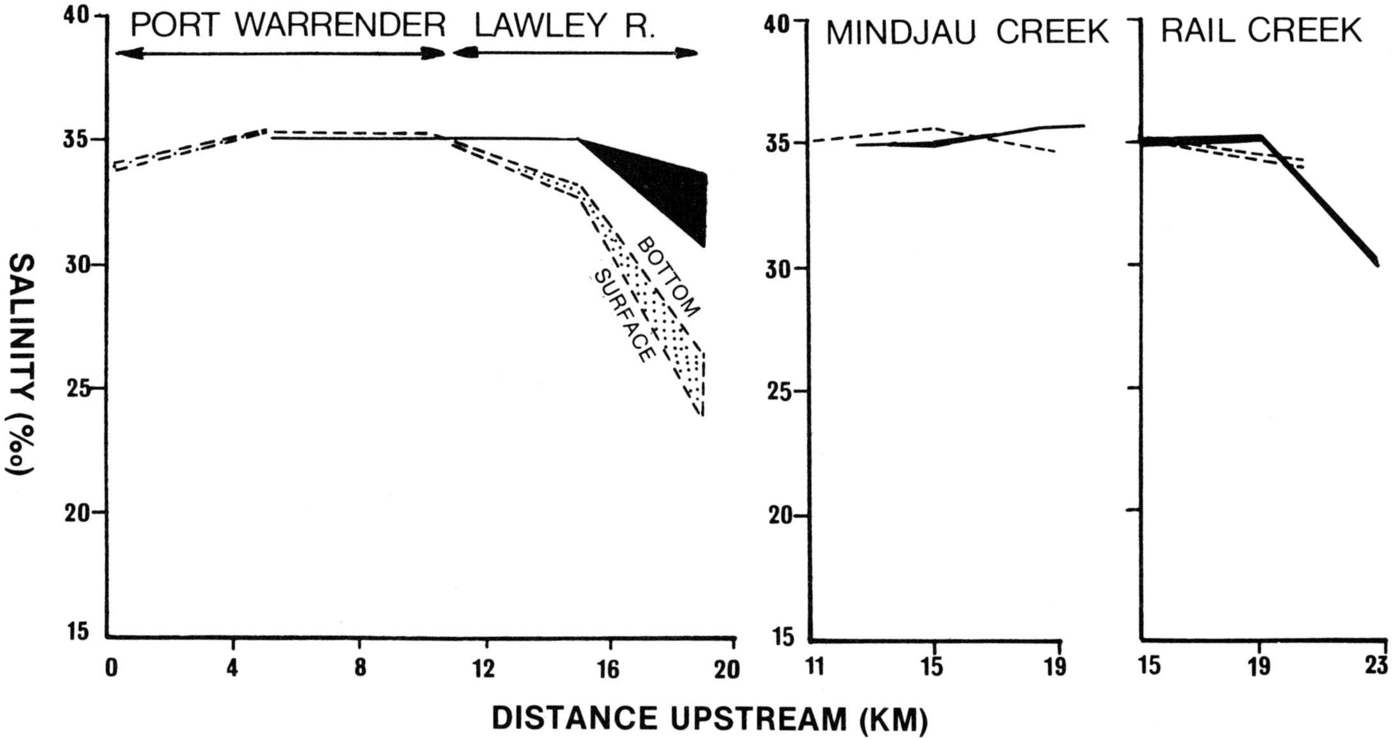

Fig. 5: High (black) and low tide (dotted) profile measurements of salinity in the Port Warrender-Lawley River system (July 1977). After Messel *et al.* 1977.

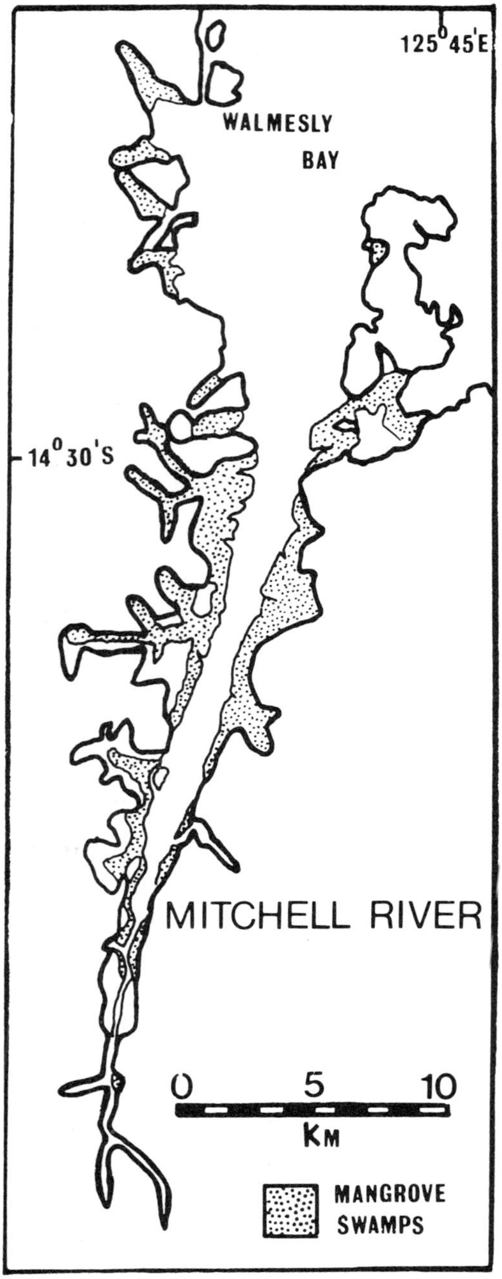

Fig. 6: Mitchell River system showing distribution of mangrove swamps.

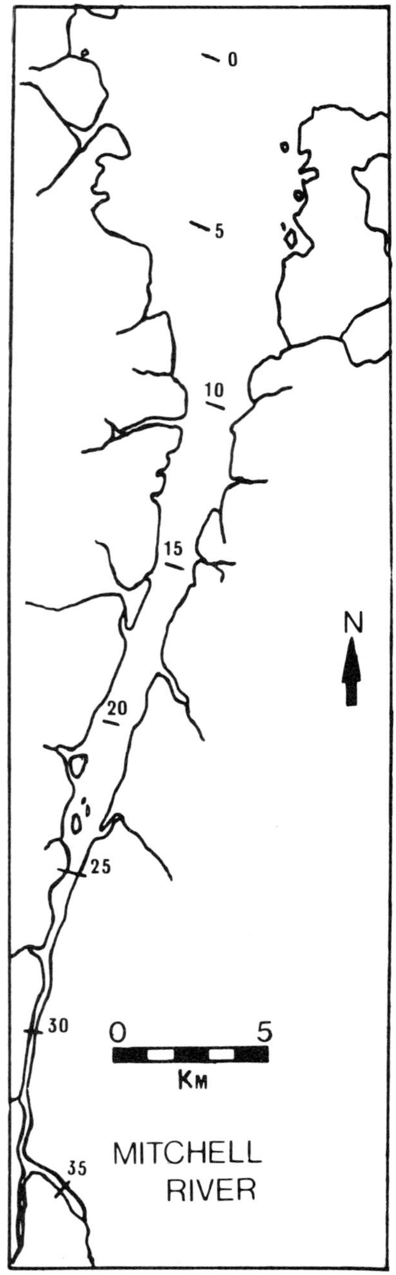

Fig. 7: Mitchell River system showing distances upstream (in km) for location of survey sites.

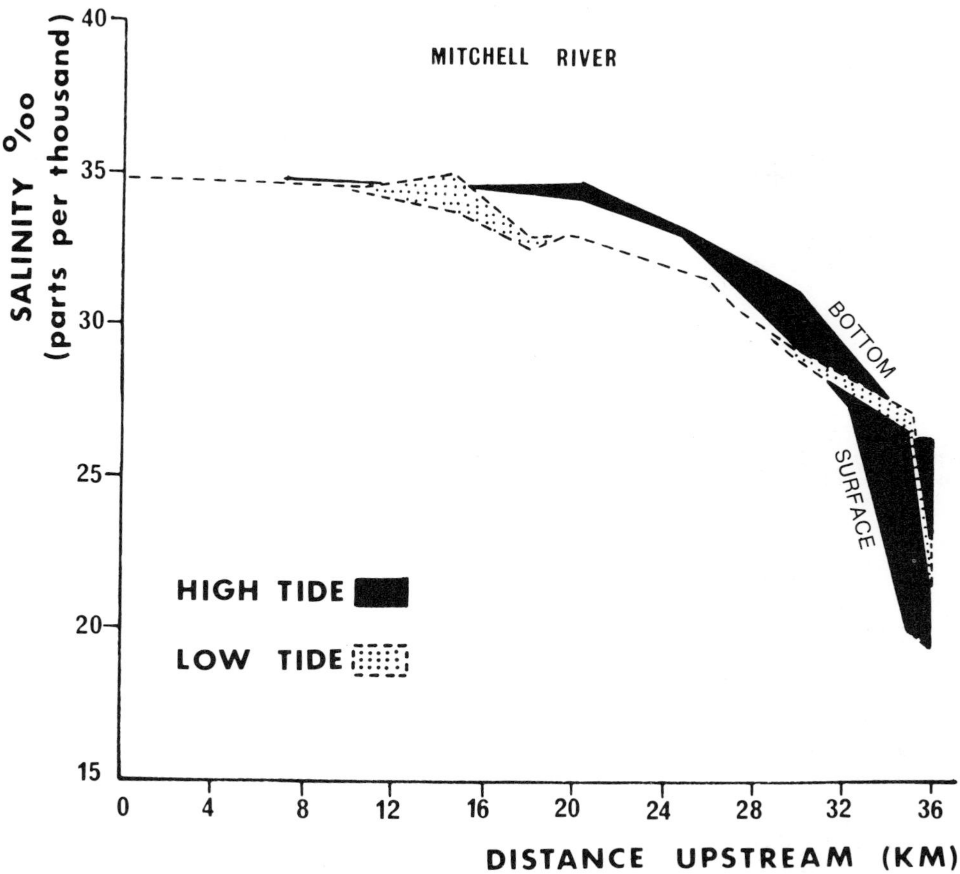

Fig. 8: High and low tide profile measurements of salinity in the Mitchell River (July 1977). After Messel *et al.* 1977.

Mitchell River System

The Mitchell River follows a fault line entering Admiralty Gulf through Walmsley Bay **(Fig. 7)**. Deeply dissected Precambrian sandstones occur along both banks. The catchment area occupies 3193 km². Mangal occupies 45 km² along the river and the short sidecreeks **(Fig. 6)**. From Pickering Point about 8 km upstream, there are a number of wide shallow embayments encompassing a further 14 km² of swamp—the dominant species here is *Rhizophora stylosa.*

Measurements of high and low tide salinity profiles **(Fig. 8)** illustrate that during July there is a decrease in salinity from 35‰ (seawater) near the river mouth to 20‰ at 36 km—the furthest limit of tidal influence.

Mangals are widespread along the sidecreeks but a relatively narrow mangrove fringe lines the main channel up to 36 km. The landward extension of the mangal is limited by rocky terrain along both banks (**Plate 2**). Extensive, gently sloping mud-banks are found throughout the entire river system, particularly along the first 20 km. The pioneer community on these mudbanks is an association of *Sonneratia alba* and *Avicennia marina* (canopy height $\leq$8 m). Low closed forest of *Rhizophora stylosa* (canopy height $\leq$8 m) occurs immediately behind this riverine fringe. The more landward forests include associations of *Camptostemon schultzii, Avicennia marina, Rhizophora stylosa* and *Xylocarpus moluccensis.*

Thickets of *Ceriops tagal* var. *australis* occur on heavy clay soils—at sites having slightly higher physiographic relief.

Distribution and assessments of cover abundance of mangrove species occurring as fringing riverside vegetation are shown in **Fig. 9**. The most abundant species are *Avicennia marina, Rhizophora stylosa* and *Aegiceras corniculatum*. Fringing forests of *Avicennia marina* and *Sonneratia alba* occur up to 26 km upstream. From this point, fringing riverside vegetation species merge into a tree association of *Avicennia marina, Rhizophora stylosa* and *Camptostemon schultzii* often with a shrub understorey of *Aegiceras corniculatum.* On rocky sites from 22.5 km upstream, an association of *Excoecaria agallocha* and *Xylocarpus moluccensis* occurs. The species *Bruguiera exaristata, Lumnitzera racemosa* and *Osbornia octodonta* occur within the river system but were not observed at any of the specified quadrats.

Roe River System

The Roe River enters Prince Frederick Harbour. On each side of the river mouth is a sheltered embayment with numerous short tidal creeks (**Fig. 10**). The catchment area of the Roe/Moran River System occupies 3570 km^2. The river course winds between Precambrian sandstone gorges—some cliffs, particularly to the north, are of basalt. The tidal range is up to 8 m and tidal influence extends 33 km upstream. Around Prince Frederick Harbour, 68 km^2 is vegetated by mangrove species. A further 25 km^2 of mangal occurs along the Roe River (**Fig. 11**). Salinity profiles upstream are shown in **Fig. 12** and are similar to other systems still receiving some fresh water input at this time of year. Several of the side creeks were also found to have significant fresh water inputs.

Distribution and assessments of cover abundance of the various mangrove species occurring as fringing riverside vegetation are shown in **Fig. 13.**

Around the river entrance and within the adjacent embayments (**Fig. 11**), *Sonneratia alba* and *Avicennia marina* form the seaward pioneer plant association. *Rhizophora stylosa* forms a band of variable width—up to 80 m—immediately behind this *Sonneratia* zone. A mixed association of *Camptostemon schultzii, Avicennia marina* and *Rhizophora stylosa* (canopy height $\leq$8 m) is present behind the *Rhizophora* zone. On more landward sites, thickets of *Ceriops tagal* var. *australis, Excoecaria agallocha* and *Avicennia marina* occur. These *Ceriops* zones are usually less than 5 m canopy height. This zone merges into the most landward mangrove community of *Avicennia marina* (canopy height $\leq$5 m). Tidal flat areas found beyond this *Avicennia* zone are in most cases devoid of any vegetative cover. Tidal flats do not occur further upstream than 17 km.

Riverside vegetation (**Fig. 13**) is a low closed forest of *Camptostemon schultzii* and *Avicennia marina* (canopy height $\leq$10 m) with a shrub understorey of *Aegiceras corniculatum* and *Aegialitis annulata.*

Although the Roe River winds through steep sandstone gorges, extensive mud-banks are present—particularly around the convex portions of the river channel. *Rhizophora stylosa* and *Sonneratia alba* are almost entirely absent from upstream riverside vegetation. An association of *Avicennia marina* and *Excoecaria agallocha* (canopy height $\leq$5 m) becomes the dominant riverside vegetation along the last 8 km of the tidal portion of the river.

Hunter River System

The Hunter River enters Prince Frederick Harbour flowing through deeply dissected sandstone cliffs. A tributary creek meets the Hunter River 4 km from its mouth (**Fig. 10**). Tidal influence extends some 15 km up both arms. The catchment area of the Hunter River System occupies some 352 km^2. The area of mangal is not extensive, being 7.5 km^2 within the Hunter River and 13.5 km^2 within 'Porosus Creek'.

Measurements of high and low tide salinity are illustrated in **Fig. 12.** The flow of fresh water entering the river system at this time of year is minimal. It is likely therefore that the Hunter River System becomes hypersaline by the end of the dry season.

Along fringing mudflats, groves of *Sonneratia alba* (canopy height $\leq$8 m) are abundant, often with a carpet-like understorey of *Aegialitis annulata* (canopy height $\geq$0.5 m) (**Plates 5** and **6**). Further upstream along both arms, this pioneer *Sonneratia* fringe merges into a mixed association of *Sonneratia alba, Camptostemon schultzii* and *Avicennia marina.* Behind this riverine fringe is a closed forest of *Rhizophora stylosa* (canopy height $\leq$8 m) (**Plate 3**). As steep rocky terrain borders the actual river channels in many places, the more landward mangrove communities are often absent. Scattered thickets of *Ceriops tagal* var. *australis* and *Excoecaria agallocha* occur on consolidated clays at a slightly higher level of physiographic relief.

Distribution and assessments of cover abundance for mangrove species occurring as fringing riverside vegetation on the Hunter River System are shown in **Fig. 14.**

It should be noted that although the species *Lumnitzera racemosa, Osbornia octodonta, Bruguiera exaristata* and *Bruguiera parviflora* were not recorded at the time in any of the quadrats, all these species occur in the river system.

Prince Regent River System

The course of the Prince Regent River follows a fault line and enters St George Basin (**Fig. 15**). The catchment area of the river system occupies 3218 km^2. The tidal range can reach 10 m and tidal influence extends 45 km upstream along the main river course. Rugged Precambrian sandstone cliffs form much of the banks, and there are few areas of mudbank along the river. Within the north bay of St George Basin, mangals occupy 70 km^2. A further 72 km^2 of mangal occurs within the southern bay of the basin (**Fig. 15**). These two bays contain the largest mangals in Western Australia. A further 24 km^2 of mangal occurs along the river and its creeks.

The low tide salinity profile (**Fig. 17**) shows that considerable fresh water was still flowing into the river system from areas of upstream drainage in July 1977. Salinity readings taken in St George Basin were as of seawater salinity (34‰).

Creekside vegetation within the north and south bays is comprised predominantly of *Camptostemon schultzii* with *Rhizophora stylosa* and *Avicennia marina* interspersed throughout (canopy height $\leq$8 m) with an inner zone of *Rhizophora stylosa*. Groves of *Sonneratia alba* (canopy height $\leq$9 m) occur nearest the seaward extremities of the bays on accreting mudbanks.

Fig. 18 illustrates the distribution and assessments of cover abundance of mangrove species occurring as fringing riverside vegetation in the Prince Regent River. Along the downstream portion of the main channel, narrow fringe forest of *Camptostemon schultzii, Rhizophora stylosa* and *Avicennia marina* (canopy height $\leq$8 m) occurs.

Plate 4 illustrates the different growth habits of *Camptostemon schultzii* on mud and rock substrate. *Sonneratia alba* is frequently observed as the pioneer species colonizing accreting mudbanks in sheltered bays along the main river course to 65 km. A shrub understorey of *Aegialitis annulata* and *Aegiceras corniculatum* is present throughout the riverside vegetation. Thickets dominated by *Ceriops tagal* var. *australis, Excoecaria agallocha* and *Avicennia marina* (canopy height $\leq$4.5 m) are limited along the river, due in part to the rocky terrain that precludes a landward extension of the mangal. *Xylocarpus moluccensis* is sparsely represented throughout the riverside vegetation while *Xylocarpus granatum* appears from 63 km upstream, forming an association with *Camptostemon schultzii* and *Avicennia marina.*

Although mangrove associations were only studied to 70 km upstream, it appears from inspection of aerial photographs that this tree association could continue to around 79 km upstream—the limit of tidal influence. *Bruguiera exaristata* and *Bruguiera parviflora* were infrequently observed. *Lumnitzera racemosa* did not occur within any of the riverside quadrats, this species normally being found in more landward mangrove associations. *Osbornia octodonta* was observed on rocky banks around the mouth of the river to 45 km upstream.

DISCUSSION

Within the study area, mangroves are present as fringing riverside vegetation along sheltered tidal waterways and within coastal embayments that provide protection from strong wave action. Species differ in their tolerance ranges for aeration and salinity. This is evident because plants are zoned in relation to frequency of tidal flooding. It is also demonstrated by the different species composition of the fringing riverside vegetation in lower and upper reaches of a river.

Pedley and Isbell (1971), Bird (1972) and Thom *et al.* (1975) have described somewhat similar mangrove habitat in northern Australia. Along the east Australian coastline, Macnae (1966) classified habitat zones using dominant mangrove species. His six zones comprised:

1.	The landward fringe	4.	*Bruguiera* forests
2.	The landward *Avicennia* zone	5.	*Rhizophora* zone
3.	*Ceriops* thickets	6.	The seaward fringe

Within the Kimberley area, all of these habitat zones occur, with the exception of zone 4—the *Bruguiera* forests. *Bruguiera* spp. were absent from nearly all quadrats. This is reflected in the fact that this area of the Western Australian coastline approaches the southernmost limit of distribution of this genus (Semeniuk *et al.*, 1978).

Thirteen species recognized as mangroves were observed in each of the river systems. An additional species, *Xylocarpus granatum*, was observed only in the upper reaches of the Prince Regent River and represents a new mangrove species recorded for Western Australia.

In many of the rivers studied, *Rhizophora stylosa* and *Sonneratia alba* are confined to sites in the outer estuaries. They are usually absent from mangrove associations occurring in the upper reaches of portions of the rivers. Thom *et al.* (1975) postulate that the distribution of *Rhizophora stylosa* could be a result of the species' inability to tolerate periodic low salinities in upstream areas. From my own examination of numerous tidal river systems in northern Australia (unpublished data), it is evident that *Rhizophora stylosa* can in fact tolerate periodic low salinities. An explanation offered here for the absence of the species from upstream sites is that sites of suitable physiographic relief for colonization are lacking. In the upper reaches of tidal portions, the most common riverside tree association is of *Avicennia marina* and *Excoecaria agallocha.* This association is normally found on more landward sites that are inundated only by spring tides. Steep rises in river and surrounding land gradients in the Kimberley result in the more landward mangrove associations now becoming the fringing riverside tree association in upstream areas. Such sites are only inundated by spring tides. *Rhizophora stylosa* normally grows in areas frequently inundated by the tide and because the upstream sites are only periodically inundated soil salinities experienced in the *Avicennia/Excoecaria* association could be well above the tolerance level of *Rhizophora stylosa* by the end of the dry season.

Sonneratia alba colonizes accreting mudbanks in the outer estuarine areas. The salinity tolerance of this species is not as wide as *Rhizophora stylosa*, but again the restricted distribution of the species is most likely a combination of lack of suitable sites for colonization and the relatively high soil salinities that would occur in the riverside *Avicennia/Excoecaria* association towards the end of the dry season.

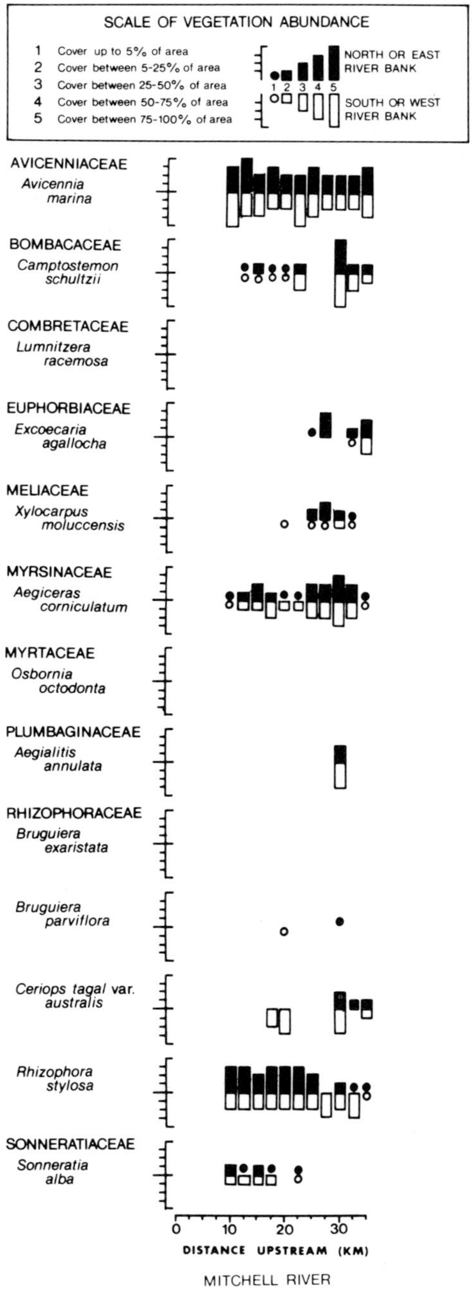

Fig. 9: Distribution and cover abundance of mangrove species within 100 m (length) x 20 m (depth) quadrats of fringing riverside vegetation along each river bank at 2.5 km intervals.

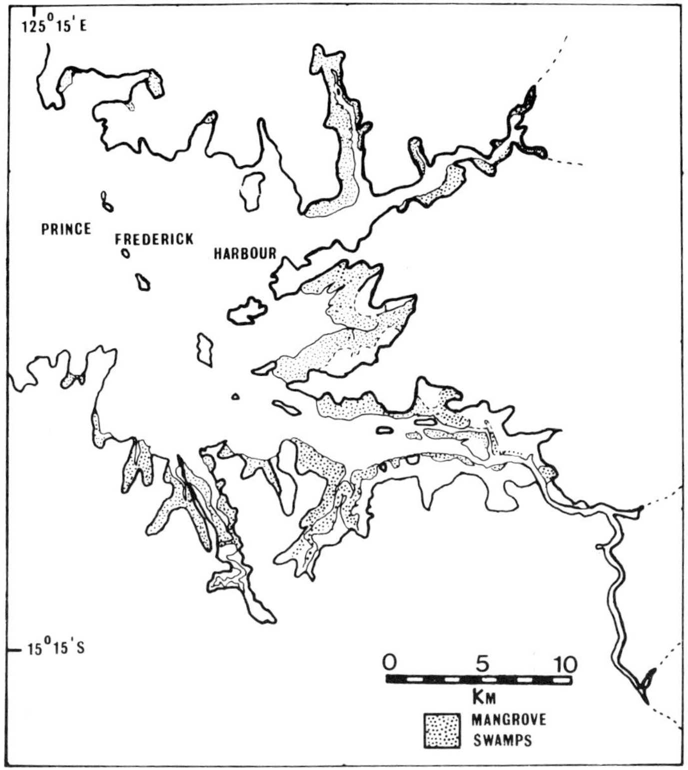

Fig. 10: Roe and Hunter River systems showing distribution of mangrove swamps.

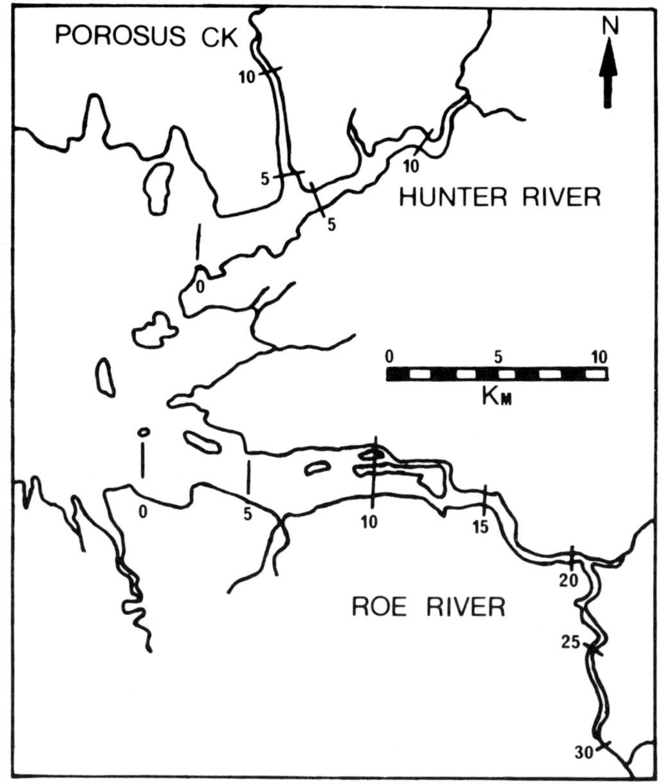

Fig. 11: Roe and Hunter River systems showing distances upstream (in km) for location of survey sites.

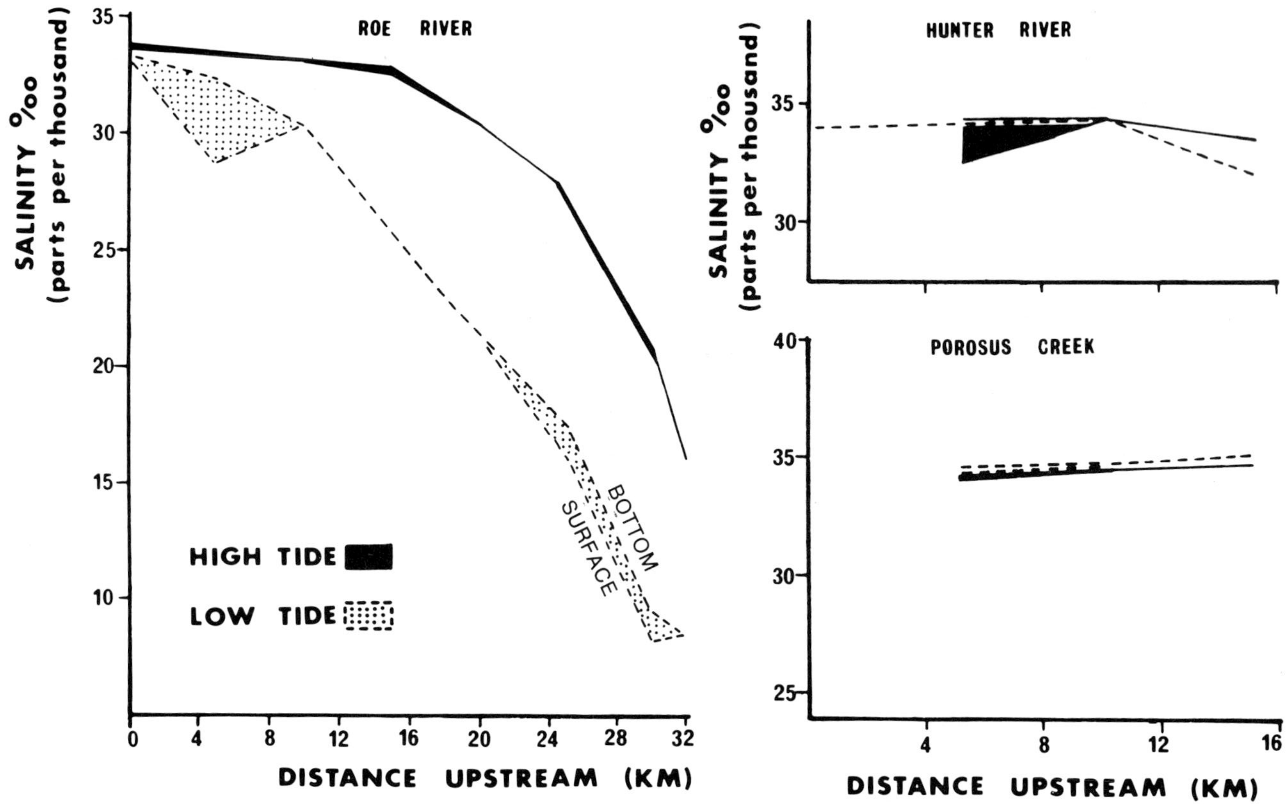

Fig. 12: High and low tide profile measurements of salinity in the Roe and Hunter River systems (July 1977). After Messel *et al.* 1977.

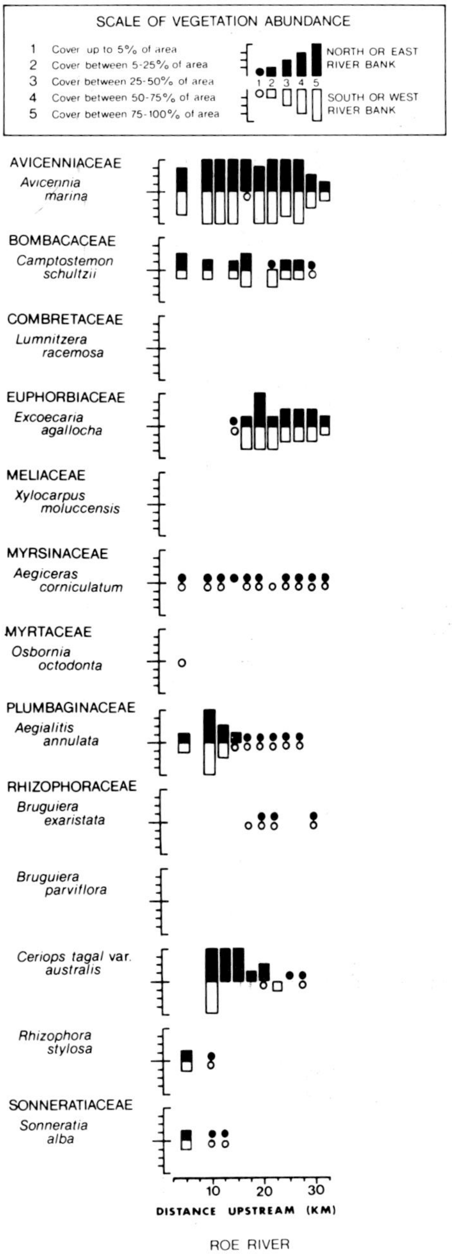

Fig. 13: Distribution and assessment of cover abundance for mangrove species within 100 m (length) x 20 m (depth) quadrats of fringing riverside vegetation along each river bank at 2.5 km intervals.

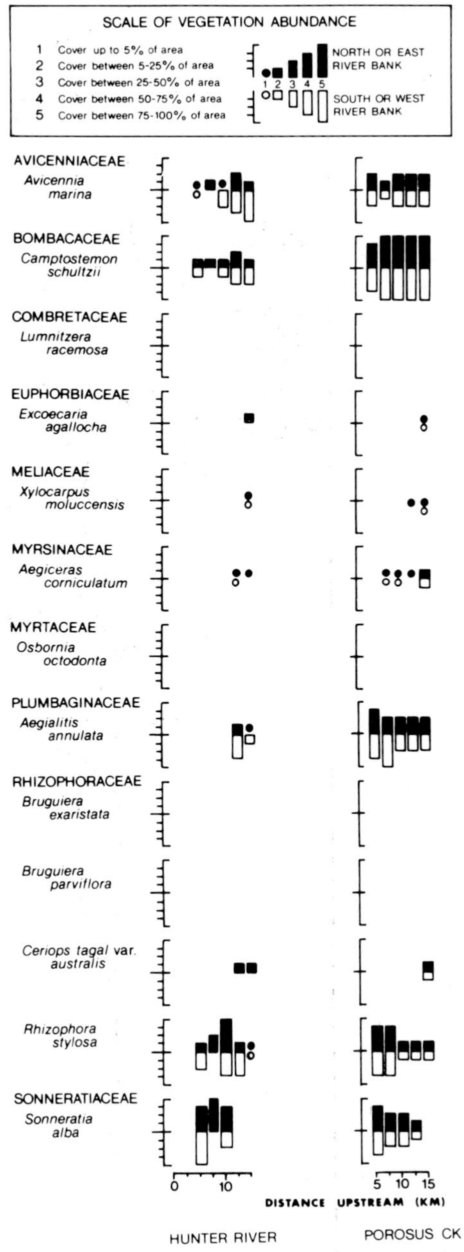

Fig. 14: Distribution and assessment of cover abundance for mangrove species within 100 m (length) x 20 m (depth) quadrats of fringing riverside vegetation along each river bank at 2.5 km intervals.

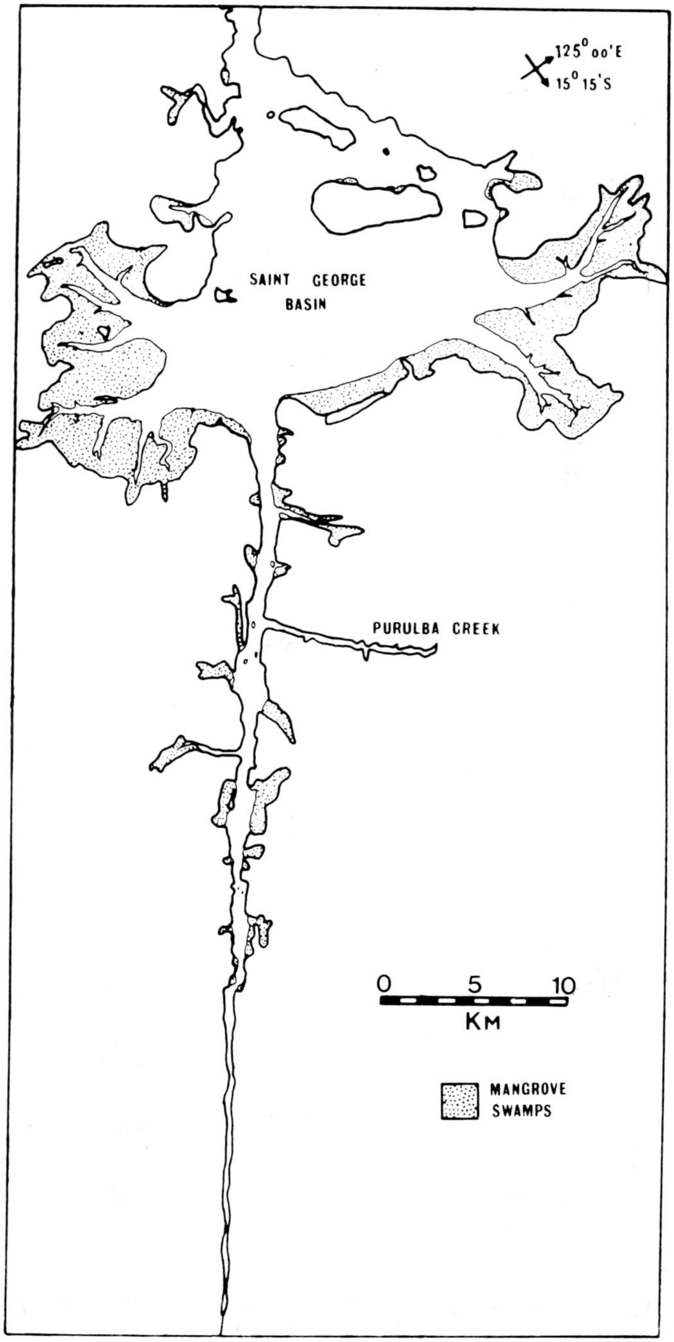

Fig. 15: Prince Regent River system showing distribution of mangrove swamps.

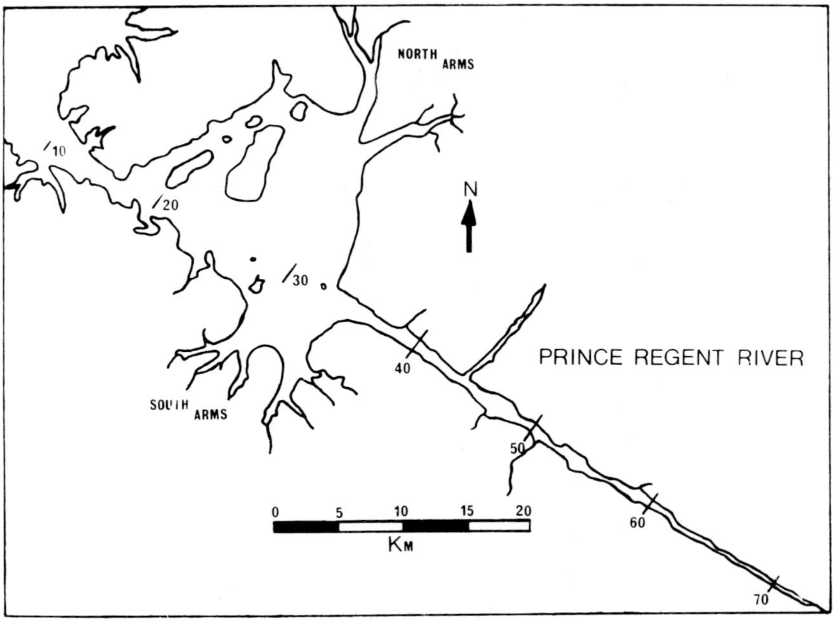

Fig. 16: Prince Regent River system showing distance upstream (in km) for location of survey sites.

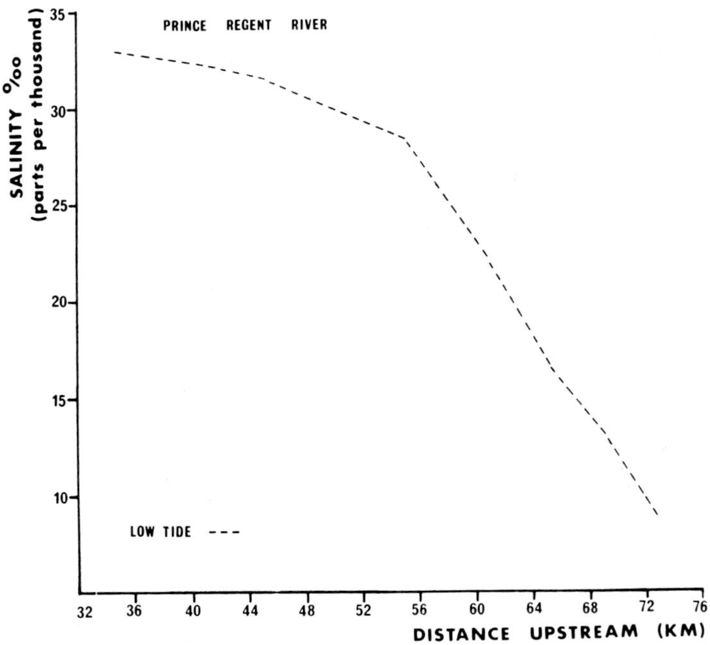

Fig. 17: Low tide profile measurements of salinity in the Prince Regent River (July 1977). After Messel *et al.* 1977.

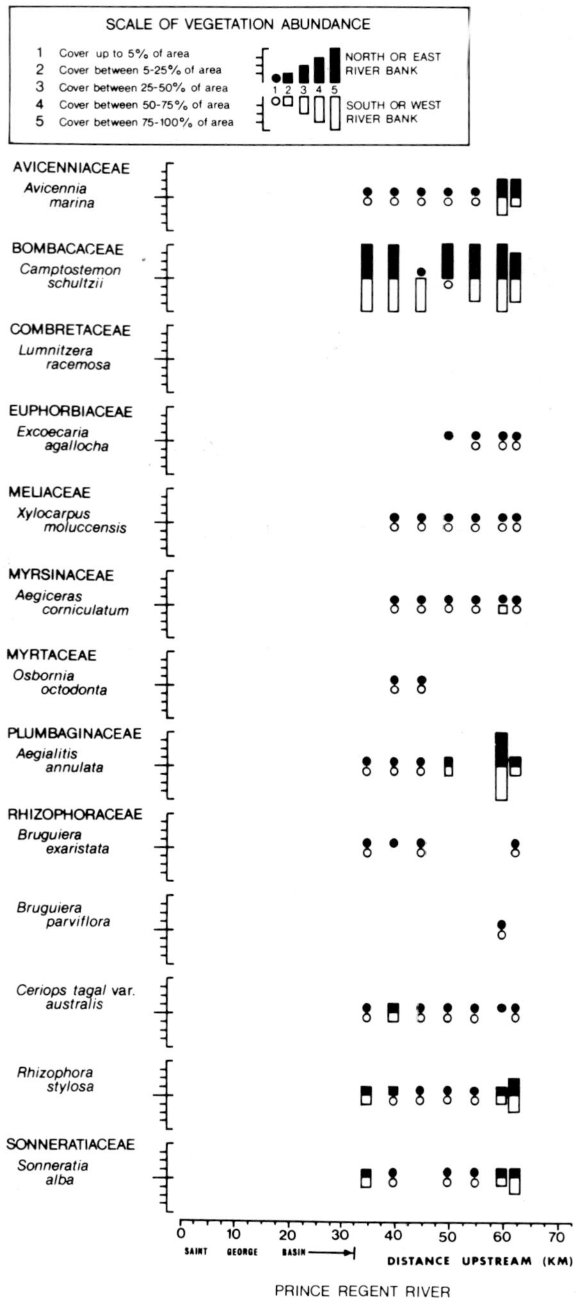

Fig. 18: Distribution and assessment of cover abundance for mangrove species within 100 m (length) x 20 m (depth) quadrats of fringing riverside vegetation along each river bank at 2.5 km intervals.

Plate 1: Association of *Avicennia marina* and *Camptostemon schultizii* colonizing low gradient mudbank near the mouth of the Lawley River.

Plate 2: *Avicennia/Excoecaria* association established in mud substrate along the sandstone banks of Surveyors Creek – a tidal arm of the Mitchell River system.

Plate 3: *Rhizophora stylosa* showing two growth habits (with and without aerial roots). *Avicennia marina* is in the centre with seedlings of *Aegialitis annulata* in the foreground. 'Porosus' Creek — Hunter River system.

Plate 4: Growth habits of *Camptostemon schultzii* on mud and rock substrates, Prince Regent River.

Plate 5: Saplings of *Sonneratia alba* with dense carpet-like understorey of *Aegialitis annulata*, Hunter River system. Note the rocky terrain limiting landward extension of the mangrove swamp.

Plate 6: Grove of *Sonneratia alba* with characteristic pneumatophores (breathing roots), Hunter River.

ACKNOWLEDGEMENTS

Thanks are due to Professor H. Messel, Head of the School of Physics, University of Sydney and to the Science Foundation for Physics within the University for provision of the considerable logistical and financial support. Thanks are also due to the crew of the R.V. *the Harry Messel* and to Dr B.R. Wilson, Western Australian Museum; P. Fuller, West Australian Wildlife Research Centre and A. Watson of the Territory Parks and Wildlife Commission for assistance in the field. Helpful suggestions in the preparation of the manuscript were given by Dr A.A. Burbidge and Dr J. Kinnear, West Australian Wildlife Research Centre, and by Dr P. Myerscough and Dr W. Magnusson, School of Biological Sciences, University of Sydney.

The study was carried out while the University of Sydney's research vessel was under contract to the West Australian Department of Fisheries and Wildlife.

REFERENCES

BEARD, J.S. (1967)—Some vegetation types of tropical Australia in relation to those of Africa and America. *J. Ecol.* **55:** 271-90.

BIRD, E.C.F. (1972)—Mangroves and coastal geomorphology in Cairns Bay, north Queensland. *J. Trop. Geog.* **35**: 11-16.

BRAUN-BLANQUET, J. (1932)—Plant sociology. Springer, Wien.

CLARK, L.D. & HANNON, N.J. (1970)—The mangrove swamp and salt-marsh communities of the Sydney District. III. Plant growth in relation to salinity and waterlogging. *J. Ecol.* **58:** 351-369.

GARDNER, C.A. (1923)—Botanical notes. Kimberley Division of Western Australia. *Bull. Forest Dep. West. Aust.* no. 32.

JONES, W.T. (1971)—The field identification and distribution of mangroves in eastern Australia. *Qd. Nat.* **20:** 35-51.

MACNAE, W. (1966)—Mangroves in eastern and southern Australia. *Aust. J. Bot.* **14:** 67-104.

MILES, J.M., KENNEALLY, K.F. & GEORGE, A.S. (1975)—The Prince Regent River Reserve environment. *Wildl. Res. Bull. West. Aust.* no. 3: 17-30.

MESSEL, H., BURBIDGE, A.A., WELLS, A.G. & GREEN, W. (1977)—The status of the salt water crocodile in some river systems of the North-West Kimberley, Western Australia. *Rep. Dep. Fish. Wildl. West. Aust.* no. 24: 1-50.

PEDLEY, L. & ISBELL, R.F. (1971)—Plant communities of Cape York Peninsula. *Proc. R. Soc. Qd* **85:** 51-74.

PERRY, R.A. (1970)—Vegetation of the Ord-Victoria area. *Land Res. Ser. C.S.I.R.O. Aust.* no. 28: 104-119.

SAUER, J.D. (1965)—Geographic reconnaissance of Western Australia seashore vegetation. *Aust. J. Bot.* **13:** 39-69.

SEMENIUK, V., KENNEALLY, K.F. & WILSON, P.G. (1978)—Mangroves of Western Australia. *Handbk West. Aust. Nat. Club* no. 12.

THOM, B.G., WRIGHT, L.D. & COLEMAN, J.M. (1975)—Mangrove ecology and deltaic-estuarine geomorphology; Cambridge Gulf-Ord River, Western Australia. *J. Ecol.* **63:** 203-232.

Received 6 February 1979 Accepted 23 October 1979 Published 28 February 1981

PART 4

OBSERVATIONS ON MAMMALS OF THE MITCHELL PLATEAU AREA, KIMBERLEY, WESTERN AUSTRALIA

D.J. KITCHENER[1], L.E. KELLER[2], A. CHAPMAN[1],
N.L. McKENZIE[3], A.N. START[4]
and K.F. KENNEALLY[5].

ABSTRACT

The Mitchell Plateau area contains a rich assemblage of mammals. Thirty-nine species were recorded, including a number of species previously poorly collected in the Kimberley. For most species, information is presented on habitat, diet and reproduction, as well as on their distribution, habitat and reproduction in the larger Kimberley region where this is thought helpful to an appreciation of the Mitchell Plateau situation.

Observations on reproduction add considerably to the small amount of information available about mammals in the Kimberley.

INTRODUCTION

Prior to the surveys of 1976-7 little was known of the mammal fauna of the Mitchell Plateau area. A.J. Marshall and J.E. Nelson, Monash University, visited the area in June 1965 and collected the Rock Rat *Zyzomys woodwardi* from deciduous vine thickets at Admiralty Gulf. In January and February 1973, L.A. Smith and R.E. Johnstone, Western Australian Museum, collected the rodent *Pseudomys delicatulus* and bats *Eptesicus pumilus, Miniopterus schreibersii, Hipposideros stenotis* and *H. ater* from sandstone sites at Crystal Creek near the Plateau. W.H. Butler visited the area in 1973 for the American Museum of Natural History. Although primarily collecting birds and reptiles he obtained the following mammals: *Pseudomys nanus* (field numbers BM 2 & 3) and *Sminthopsis* cf. *douglasi* (BM 4) from riverine sites at Camp Creek, near our line 11 **(Fig. 1)** (29-30 July); *Miniopterus schreibersii* (BM 5,10,18-43), *Macroglossus minimus* (BM 8-10), *Eptesicus pumilus* (BM 6,7,12-17, 44-66) from sandstone sites at Crystal Creek, Crystal Head, near line 53 (7-10 August); and *Miniopterus schreibersii* (BM 74) from plateau slope sites at Port Warrender, near line 51 (23 October). This 1973 collection is lodged in the American Museum of Natural History.

1 Western Australian Museum, Francis Street, Perth, Western Australia. 6000
2 Field Museum of Natural History, Roosevelt Road, Chicago, Illinois, 60605, U.S.A.
3 Western Australian Wildlife Research Centre, Mullaloo Drive, Wanneroo, Western Australia. 6065.
4 National Parks Authority, Hackett Drive, Crawley, Western Australia. 6009.
5 Western Australian Herbarium, George Street, South Perth, Western Australia. 6151.

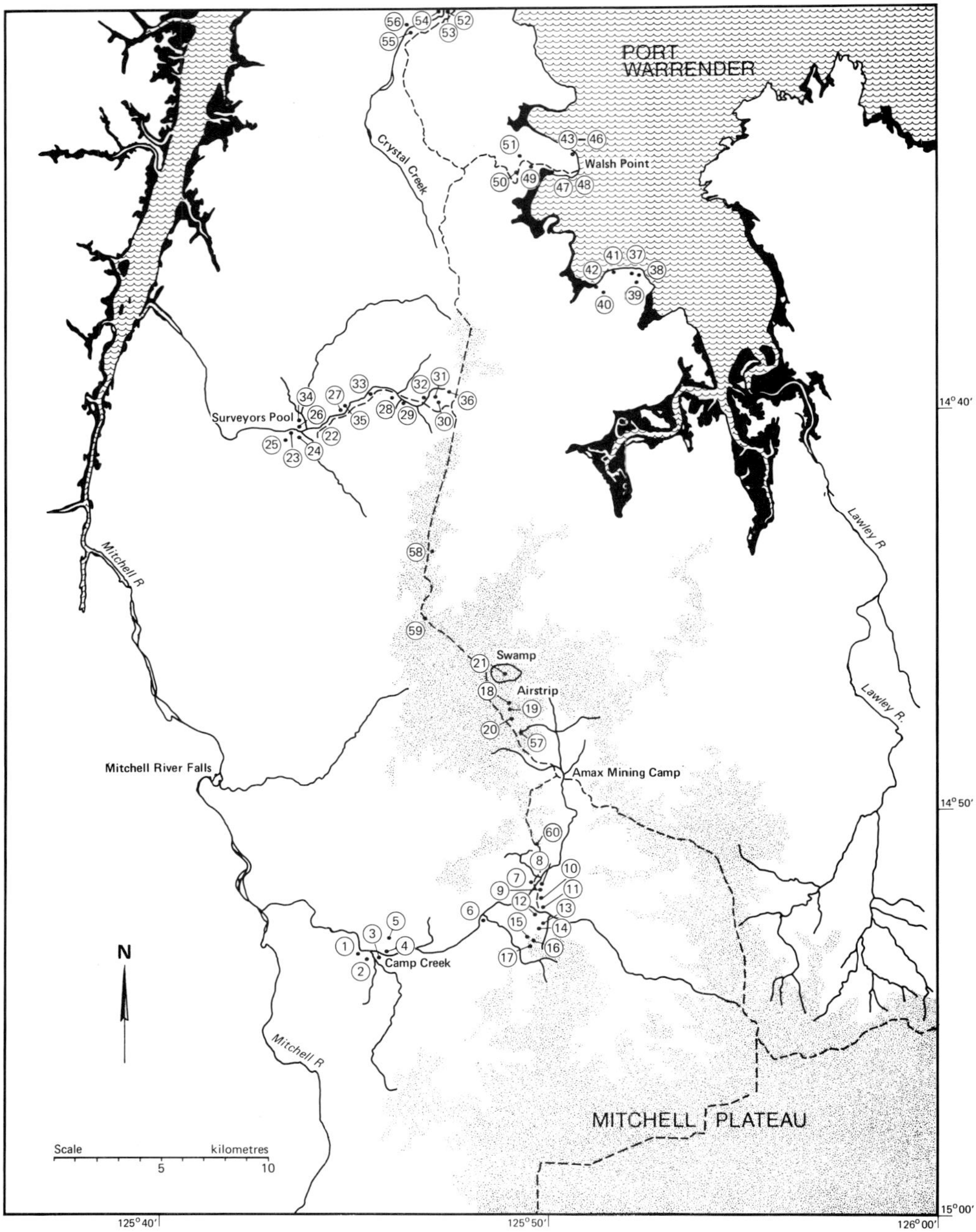

Fig. 1: Location of traplines (1-60) established at Mitchell Plateau in 1976-1977. The vegetation and substrate descriptions of these lines are in **Appendix II.**

Three expeditions to the Mitchell Plateau area are reported in this paper. Work in June 1976 was primarily concerned with vegetation, although a certain amount of mammal collecting was undertaken. Expeditions in October-November 1976 and March 1977 aimed to collect mammals in each of the major habitats, including deciduous vine thickets which are rare, confined to the north-west Kimberley coast, and previously little studied. The other broad habitats recognised by us, along with the general climate and geology and detailed vegetation description of the surveyed area, are described in Wilson, Hnatiuk and Kenneally, Johnstone & Smith (Parts 1,2,5, this publication) respectively. It was hoped that by trapping at the end of the dry season (October-November) and again at the end of the wet season (March), information would be gained on mammal population status and habitat utilisation.

Reproductive information obtained from Mitchell Plateau specimens is presented in conjunction with similar information obtained from other female Kimberley specimens in Western Australian Museum collections and from pertinent literature. This enables some generalisations on the period of birth in some groups of Kimberley mammals.

Distributional records on Mitchell Plateau mammals for the wider Kimberley area are also presented, to provide an appreciation of the Mitchell Plateau mammal assemblage in the Kimberley.

METHODS

Mammals were collected by spotlighting and shooting at night, by mist-netting in mangroves or over pools and creeks, and by trapping. A variety of traps was employed

(i) *large Breakbacks* — a commercial rat trap,

(ii) *small Breakbacks* (Museum Specials) — about the size of a commercial mouse trap,

(iii) *Victor Rat traps* — intermediate in size between (i) and (ii) but differing in that the release mechanism is operated by horizontal as well as vertical movement of the bait attachment,

(iv) *small Elliott traps* — collapsible live traps with solid aluminium walls and dimensions ca. 10x10x40 cm,

(v) *large Elliott traps* — similar to (iv) but with dimensions of ca. 20x15x50 cm,

(vi) *Sherman traps* — similar to (iv) but with dimensions of 7x8x20 cm,

(vii) *Cage traps* — a collapsible trap made from wire mesh with dimensions of 60x50x60 cm.

These traps were kept baited with 'universal bait' — a mixture of peanut butter, bacon, raisins and oats. They were checked each morning.

Weight of all mammals collected and body measurements from all skinned specimens were recorded in the field during the October-November 1976 and March 1977 surveys. Unless otherwise stated, body measurements presented in this paper were recorded in the laboratory from alcohol-stored specimens. The abbreviations for measurements given in this text are as follows:

W = weight, **H-V** = head to vent length, **T-V** = tail to vent length, **HF** = hind foot length, **FA** = forearm length, **Ti** = tibia length, and **E** = ear length, **N** = sample size, **NA** = measurements not available.

Most specimens were fixed in 10% formalin then preserved in 75% ethyl alcohol. All specimens were dissected in the laboratory and their reproductive organs examined *in situ*. Stomach contents of specimens collected at Mitchell Plateau in October-November 1976 were identified when present. Histological preparation (using haemotoxylin and eosin) was carried out on the reproductive tract of selected specimens of *Zyzomys argurus, Z. woodwardi, Conilurus penicillatus, Macroglossus minimus* and *Mesembriomys macrurus.*

Specimens from all 3 Mitchell Plateau surveys were registered into the Western Australian Museum collections. Those collected have the following catalogue numbers:—

June 1976: M14565-80
October-November 1976: M15501-M15880
March 1977: M15211-M15356

Specimens from other Kimberley localities referred to in this report, from the combined Western Australian Museum and American Museum of Natural History collection made by W.H. Butler in 1963-5, have **B** prefixed numbers.

The terminology used to describe vegetation other than deciduous vine thickets or mangroves is based on that of Muir (1977); his classification is reproduced in **Appendix I**. Where sufficient information is recorded on vegetation, foliage cover and life form to enable a precise Muir classification to be used, the vegetation description is capitalised, e.g. Open Low Woodland. Where this information is not available, or a grouping of several Muir categories is favoured, a general term is used and this is not capitalised, e.g. woodland.

June 1976 Survey

Between 12-26 June 1976, K.F. Kenneally studied the vegetation of the Mitchell Plateau area. While there, he also collected mammals by shooting and trapping. His traplines consisted of 10 traps, small and large Elliott traps set alternately at 40 m intervals. These lines were located in the Walsh Point and Mining Camp areas and can be related to the following lines set in October-November 1976: line 21 (40 trapnights*), adjacent to lines 50 (40 trapnights), line 47 (30 trapnights), adjacent to line 48 (30 trapnights).

October-November 1976 Survey

Between 16 October and 6 November the mammalogy group operated in three teams:

team 1: D.J. Kitchener & D.T. Kitchener
team 2: A. Chapman & N.L. McKenzie
team 3: L.E. Keller, Mr & Mrs W.S. Street, M. Ellis & A.N. Start

*A trapnight is one trap set for 24 hours.

Traps were set in lines at 54 localities shown in **Fig. 1**. Description of the habitat at each line is presented in **Appendix II**. The trapping schedule for each team was:

team 1: lines 9, 13-16 (18-24 October), 10-12 (19-24 October), 43-46 (27-30 October), 52-54 (1-4 November).

team 2: lines 22-24 (17-22 October), 25-26 (18-23 October), 27 (18-27 October), 28-29 (19-24 October), 30-31 (19-27 October), 32-33 (20-25 October), 34 (23-27 October), 35-36 (24-27 October), 37-38 (29 October-2 November), 39-40 (29 October-1 November), 41-42 (30 October-2 November).

team 3: lines 1-8 (18-23 October), 18-21 (24-29 October), 47-51 (27 October-5 November).

This trapping resulted in a total of 7138 small mammal trapnights (from Elliotts, Breakback, Sherman, Victor Rat and Museum Specials) and 575 large mammal-trapnights (from cage traps).

Unfortunately, because of the limited number of different traps available, the teams did not trap in the same manner. **Team 1** traplines consisted of 20 traps set at 10 m intervals, alternating a small Elliott with a large Breakback trap. A Cage trap was set near the first and last trap in each line. **Team 2** placed similar lines except that lines 22-36 had only 10 small mammal traps and 1 cage trap. **Team 3** traplines consisted of 8 stations placed 20 m apart. Each station had a Sherman, Victor Rat Trap, and 2 Museum Specials; a large Cage trap was placed at the second and fourth station.

March 1977 Survey

A. Chapman, L.E. Keller and J. Henry collected mammals in the area between 11-24 March 1977. They were unable to reach many of the trapline sites used by previous expeditions because flooding made some areas inaccessible. They placed 11 traplines:— lines 7, 43, 48, 49 and 51, just as in October-November 1976, and lines 55-60, in habitats similar to those trapped in October-November 1976 but at new localities **(Fig. 1)**. Trapping procedure was that of **Team 3** in the October-November 1976 survey. The location of these lines is given in **Fig. 1**. The coastal lines (43, 48, 49, 51, 55 & 56) were trapped between 12-17 March and lines on the lateritic plateau (7, 57, 58, 59 & 60) between 19-23 March. This trapping resulted in 1888 small mammal-trapnights and 118 large mammal-trapnights.

ANNOTATED SPECIES LIST

MACROPODIDAE

Macropus robustus Gould, 1841 **Euro**

A female with pouch young (M15833-4) was shot at 1500 hrs on 26 October 1976, 3 km E of trapline 22 in Low Open Eucalypt Woodland on volcanics. The pouch young weighed 1180.0 gm. Another adult was sighted at 1200 hrs at line 3 in Open Low Woodland of *Planchonia* over Dense Tall Grass on volcanics. A skull was found at line 53, and 2 at line 1 (M15847-49).

At Mitchell Plateau the Euro was seen less often than *M. antilopinus*. It was not observed on the lateritic plateau by any expeditions to the area. The species is widely distributed throughout the Kimberley.

Body measurement (field recorded):
1♀ W: N.A.; H-V: 840 mm; T-V: 700 mm; HF: 280 mm; FA: 165 mm; E: 100 mm.

Macropus antilopinus (Gould, 1842) **Antelope Kangaroo**

In October-November 1976 an adult male (M15837) was shot at 2100 hrs in Open Low Woodland of *Eucalyptus* spp. over spinifex on sandstone near line 1. A female (M15838) was shot 6 km W. Mining Camp in *Livistona* Low Woodland B on laterite; another female (M15839) was shot 400 m N Crystal Creek Falls. A skull was found near line 1 (M15845) and near line 42 (M15846). Two adults (sex undetermined) were seen feeding at 2100 hrs on low grass in a damp (sandstone) watercourse of Camp Creek near line 1. Single females, each with a small young at heel, were seen at about 2100 hrs in Low Woodland A of *Eucalyptus* sp. near line 7 (on volcanics) and 53 (on sandstone).

On 19 January 1973, L.A. Smith saw a female with a furred pouch young with its head out of the pouch, in woodland on the lateritic plateau near the coast.

On 17 June 1976 an adult female with a small unfurred pouch young with eyes closed (M14570-1) was shot at 1730 hrs on the western edge of the lateritic plateau in *Eucalyptus* woodland.

The Antelope Kangaroo was not seen on the lateritic plateau by L.A. Smith and R.E. Johnstone between 18-24 June 1975 or by the expeditions in October-November 1976 and March 1977. They were, however, occasionally seen there by Smith and Johnstone during their January-February expedition to the area. L.A. Smith writes in his field book that a group of 2-4 individuals was frequently seen on the lateritic plateau at a site "where (it) was constricted to a narrow neck where water and good grazing were in close proximity".

Although the Western Australian Museum collections have specimens of *M. antilopinus* from only a few Kimberley localities (Beverley Springs, Mt Hart Homestead, Parrys Creek and Negri River) it probably occurs throughout the Kimberley. Frith and Calaby (1969) state that the Antelope Kangaroo is distributed more or less continuously through tropical eucalypt woodland but shows a preference for areas "where low ridges separate lightly timbered grassy glades."

Body measurements (field recorded):
1♂ and 1♀ respectively: W: NA; H-V: 800, 760 mm; T-V: 770, 740 mm; HF: 264, 270 mm; FA: 190, 165 mm.

Macropus agilis (Gould, 1842) **Sandy Wallaby**

A female with pouch young (M15835-6) was shot at 1200 hrs on 20 October 1976 ca. 1 km E. Surveyors Pool in Open Low Woodland *Eucalyptus* sp. over cane grass on black soil. A skull (M15843) was collected from a Bower Bird's nest near line 35, and a lower mandible (M15844) from the ground near line 7. An adult male was found dead and floating in a lagoon on Camp Creek near line 7. The Sandy Wallaby was not seen on the lateritic plateau on any of the surveys; in this area it appears to be confined to country between the laterites and sandstone.

The Sandy Wallaby occurs throughout the Kimberley but Gooding (1963) states that they are mostly "along the river frontages of the Fitzroy, Lennard and Ord Rivers where they sometimes reach plague proportions."

Body measurements (field recorded):
1♀ W: 7600 gm; H-V: 700 mm; T-V: 690 mm; HF: 230 mm; FA: 190 mm; E: 80 mm.

Petrogale brachyotis Gould, 1841 **Short-eared Rock Wallaby**

In November 1976 a skull and parts of a skeleton (M15850-1) were found near line 38, on a ledge on cliffs of Carson Volcanics above deciduous vine thicket. One live animal was seen in the same thicket.

An adult male (M15355) was shot at Crystal Creek in March 1977 at 1900 hrs in low scrub over rugged sandstone at line 52.

This Rock Wallaby is widespread and fairly common throughout the Kimberley. The Western Australian Museum collections have specimens from the following Kimberley localities: Kunmunya Mission, Wotjulum, Kalumburu, Ord River area, Prince Regent River Reserve, Drysdale River National Park and Napier Downs Homestead.

Body measurements (field recorded):
1♂ W: 4200 gm; H-V: 510 mm; T-V: 460 mm; HF: 135 mm; FA: 85 mm; E: 48 mm.

Petrogale burbidgei Kitchener & Sanson, 1978 **Warabi**

Between 22 October and 3 November 1976, 3 adult females, each with a pouch young (M15827-32) were collected near line 24 and near line 53. The first locality was in an Open Low Woodland A of *Terminalia* sp. and *Eucalyptus brachyandra* over Open Hummock Grass on sandstone; the other was on rugged sandstone cliffs with occasional shrubs including *Ficus* sp. and *Acacia* sp. over Open Hummock Grass overhanging Crystal Creek. Both these localities were on King Leopold Sandstone.

This species is known only from Mitchell Plateau, Prince Regent River Reserve and several islands in the Bonaparte Archipelago. It appears to favour low woodlands including such species as *Eucalyptus* spp., *Owenia vernicosa, Terminalia* sp. and *Ficus* sp. over spinifex on rugged King Leopold sandstone.

The 3 pouch young collected at Mitchell Plateau in mid-October or early November weighed 51.0, 77.0 and 85.0 gm. An additional female from Bigge Island collected on 3 June 1972 had a pouch young weighing 18.5 gm; 1 from Prince Regent River Reserve in August 1975 had a pouch young of 48.0 gm. Detailed measurements present in Kitchener & Sanson (1978).

PHALANGERIDAE

Trichosurus arnhemensis Collett, 1897 **Northern Brush Possum**

A male (M15348) was shot at 2030 hrs in a *Eucalyptus miniata*, near trapline 58 on 12 March 1977 in a Low Woodland A of *Eucalyptus* sp. over a Low Woodland B of *Livistona eastonii* over cane grass on laterite.

This species is recorded in Western Australian Museum collections from only a few localities in the Kimberley (Drysdale River area, Kalumburu, Halls Creek and Myroodah) and there is no habitat information with these records. Calaby and Keith (1974) report it to be common in the Port Essington area, Northern Territory, where it inhabits eucalypt forest and woodland and the edge of monsoon forest. Johnson (1964) records *Trichosurus vulpecula arnhemensis* as present in low numbers throughout coastal Arnhem Land, including the larger nearby islands.

Body measurements (field recorded):

♂ W: 1500 gm; H-V: 380 mm; T-V: 240 mm; HF: 560 mm; FA: 690 mm; E: 54 mm.

Petaurus breviceps (Waterhouse, 1839) **Sugar Glider**

A male (M15790) was shot in October 1976 at 2115 hrs, 3 km SE of trapline 52. It was in a eucalypt tree in Low Woodland B of *Eucalyptus* spp. and *Melaleuca leucadendron* on black soil.

A male (M15347) was shot in a *Erythrophleum chlorostachys* tree at 2010 hrs, 2 km E of line 50; it was in Low Open Woodland A over a Tall Grass of *Chrysopogon latifolius* on volcanic scree.

This species is widespread in the more northern parts of the Kimberley in woodlands of *Eucalyptus* spp., *Acacia* spp. and *Owenia* sp. over grassland. (See also McKenzie *et al* 1977, 1978). In the Northern Territory, Johnson (1964) reports it abundant on Groote Eylandt both in "eucalypt forest" and on "flats in wattle scrub on sand dunes". Calaby & Keith (1974) obtained their only specimen from Cobourg Peninsula in "eucalypt forest".

The Western Australian Museum collections have several females and young of this species from the Kimberley. In September 1955, 2♀ (M3107-8) from Wotjulum (16°11′S, 123°37′E) were not pregnant but had a pouch with very deep lateral pockets and elongate teats and were probably lactating when collected. On 13 September 1975 a female (M13948) was collected near Mt Heytesbury. It had 2 partially furred pouch young with crown to rump length of *ca* 45.0 mm.

Body measurements (field recorded):

2♂ W: 118, 135 gm; H-V: 169, 182 mm; T-V: 215, 21 mm; FA: 51, 51 mm; HF: 26.6, 28.7 mm; E: 23.8, 24.1 mm.

Wyulda squamicaudata Alexander, 1918 **Scaly-tailed Possum**

Part of a mandible (M18099) was found by W.H. Butler in November 1978 adjacent to Mitchell River at 14°35′00″S, 125°43′15″E. It was lying on King Leopold Sandstone vegetated with sparse *Plectrachne* and *Acacia*. This species, although infrequently recorded, has a reasonably wide distribution within the Kimberley and has been collected from the following localities: Violet Valley Homestead, Wotjulum, Kalumburu, Boongaree I. and Prince Regent River Reserve.

PERAMELIDAE

Isoodon macrourus (Gould, 1842) **Brindled Bandicoot**

Nine (M15819-26 and M15874), including 3 pick-up skulls, were collected in October 1976. Trapped specimens were from the following lines: 25 (1 ♂), 27 (1 ♂), 30 (1 ♀), 31 (1 ♀), 48 (1 ♀). A male was shot adjacent to line 27 in Open Woodland of *Eucalyptus* spp. with occasional *Pandanus* over *Sorghum* sp. on black soil plain. Pick-up skulls were collected from near line 28, and at line 31 and 32. Others were released from the following lines: 25 (1 ♀, 585 gm), 27 (2 ♂, both 625 gm) and line 30 (1 ♀, not weighed).

These records (excluding pick-up material) were from a range of habitat encompassing: riparian woodland of *Eucalyptus* spp. over *Melaleuca viridiflora* shrubs and tall grasses on volcanics; deciduous vine thicket at the western edge of the lateritic plateau; Woodland of *Eucalyptus* and *Terminalia* over Dwarf Scrub C, over Very Open Tall Grass and occasional spinifex on flat sandstone; Low Woodland A of *Eucalyptus apodophylla* over Low Forest B, over Tall Grass on black soil plain; Low Forest A of *Eucalyptus confertiflora* over Dwarf Scrub D on red loam soil on edge of lateritic plateau. Several were also sighted in this last habitat and on the lateritic plateau in Low Woodland A of *Eucalyptus miniata* and *E. tetrodonta* and *Livistona eastonii.*

McKenzie *et al.* (1975) captured specimens in riparian vegetation on deep alluvial soils on both sandstone and volcanics in the Prince Regent River Reserve. The species has been recorded from a few widely separated localities in the north and north-western Kimberley (Wotjulum, Kalumburu, Blythe Creek, Prince Regent River Reserve, St Andrews I. and Camden Harbour).

At Mitchell Plateau in October none of the females were pregnant or lactating. The Western Australian Museum has several other female alcohol-stored specimens collected in August 1974 (Prince Regent River Reserve, M12381-2 and M12385-6) and on 3 July 1960 (Kalumburu, M4092). None of these are pregnant or lactating.

Stomachs contained Orthoptera (Acrididae), Blattodea, Hemiptera (Cicadidae), Coleoptera (Scarabidae, Carabidae), Hymenoptera (Symphyta), numerous Lepidoptera larvae and Araneida.

Mitchell Plateau specimens appear to have a shorter body and tail than those from Northern Territory reported in Calaby and Keith (1974) and Johnson (1964).

Body measurements (mean and range):

	N	W (gm)	H-V (mm)	T-V (mm)	HF (mm)	FA (mm)	E (mm)
♀ ♀ (adult)	3	600 (590-610)	276 (260-289)	141 (132-146)	5.6 (5.9-6.0)	4.5 (4.4-4.6)	3.3 (3.2-3.4)
♂ (adult)	1	1000	311	140	6.7	5.7	3.5

DASYURIDAE

Dasyurus hallucatus Gould, 1842 **Little Northern Native Cat**

In June 1976, 2 (M14565-6) were collected from line 47 (2♀) and 6 were released from line 47.

In October and November 1976, 10 (M15807-14 and M15816-18) were collected from the following lines: 1 (2♂, 2♀), 22 (1♀), 32 (1♀), 37 (1♀), 39 (1♀), 41 (1♀), 43 (1♂) and 44 (1♂). Another female (M15815) was shot 2 km E line 53 at 2145 hrs in Open Low Woodland over burnt grassland on volcanics; 2 males were released from line 25 (362 and 365 gms), one male from line 34 (410 gms), and one unsexed animal was released from line 43. One was sighted at 2100 hrs on sandstone near line 53.

In March 1977 a female (M15349-50) was collected from line 43 and 56.

In June 1976, these animals were trapped from *Eucalyptus miniata, E. tetrodonta* Open Low Woodland on the edge of a watercourse. In October-November 1976, proportional to trapping effort, most were captured in sandstone country (6 specimens, 76 cage trapnights) with the upper vegetation strata variously of Hummock Grass, Open Low Scrub, Low Woodland A, and Woodland. They were also common in the deciduous vine thickets on volcanics and laterites (5 specimens, 177 cage trapnights) but were less frequently collected in other vegetation types on volcanics (4 specimens, 242 cage trapnights). A single animal was trapped in Open Low Woodland B of *Terminalia canescens* and *Pandanus spiralis* fringing a watercourse, and another from Low Forest A of *Eucalyptus confertiflora* on the edge of the lateritic plateau. In March 1977 a specimen was collected from both deciduous vine thicket on volcanics and from Low Woodland A of *Eucalyptus* spp. and *Owenia vernicosa* on rugged sandstone.

These observations and those of McKenzie *et al.* (1975) and McKenzie *et al.* (1978) confirm Johnson's (1964) statement that this species prefers rocky habitat but is not necessarily restricted to it. The species appears to have a patchy distribution throughout the Kimberley.

At Mitchell Plateau, neither female collected on 28 June 1976 was pregnant; their teats were small and pouch area moderately well developed. Two females collected on 19 October 1976 were carrying young (2 and 8), the young of both these litters were small, partially furred, and had hind foot lengths of 8.2 to 10.8 mm and 5.5 to 6.5 mm respectively. Six other females collected between 21 October-2 November 1976 were not pregnant and had no pouch young; they appeared to have been recently nursing young because their pouches, mammary glands, and teats were considerably enlarged. Between 13-15 March 1977, 2 juvenile females were trapped. These were about three-quarters adult size and had hind foot lengths of 36.8 and 37.0 mm.

The Western Australian Museum has the following female specimens from the Kimberley: 1-7 June 1972 (Bigge Island, M9308, M9310); 17 June 1965 (Inglis Gap, M7164); 1-4 July 1973 (Anjo Point, M10355; Boongaree I., M10358, M10360); 3 July 1960 (Kalumburu, M10643); 26 August 1975 (Prince Regent River Reserve, M12411) and 14 December 1965-20 January 1966 (Kalumburu, M7165-6, M7168,

M7171, B1869, B1878 & B1932). None of the June or July specimens showed indications of recently carrying or nursing young; those dissected, except M7164 which was a skin and skeleton only, were not pregnant. The female collected in late August from Prince Regent River Reserve had 8 tiny pouch young with hind foot length ca. 2.2 mm. The single pouch young collected on 30 October 1966 (M10606) at Koolan I. was furred and had hind foot length of 16.0 mm. Between 14 December 1965 and 20 January 1966, W.H. Butler collected 7 females at Kalumburu. In his field notes he does not record any of these as carrying pouch young. One spirit specimen and 3 skins from Butler's collection were examined. The spirit specimen had only slight pouch development, tiny teats, and was not pregnant. The skins all had enlarged pouches and elongate teats suggesting that they had recently nursed young. This is supported by the collection of 3 juveniles (trapped singly) at Kalumburu between 10-17 January 1966. These juveniles were about half adult size and had hind foot lengths of 20, 22 and 28 mm.

Combined, these observations suggest that in the north and north-west Kimberley *D. hallucatus* gives birth to young during the latter half of the dry season (August to November). Most young would appear to have left the pouch before the middle of the wet season (January). Johnson (1964) and Calaby and Keith (1974) report that in the Northern Territory the breeding season of *D. hallucatus* also seems to be during the dry season.

The stomach of one specimen contained Tettigoniidae, Mantidae (egg cases), Gryllidae, Chilopoda, Coleoptera, Araneida and unrecognizable organic material; the stomachs of 7 others were empty.

Mitchell Plateau specimens fall within the range of body size for this species in the Northern Territory as presented in Calaby & Keith (1974) and Johnson (1964).

Body measurements (field recorded) (mean and range):

	W (gm)	H-V (mm)	T-V (mm)	HF (mm)	FA (mm)	E (mm)
♀♀ (parous only)	323.5 (296-355) N=8	243 (216-292) N=7	215 (204-231) N=6	38.2 (36.6-39.8) N=8	44.9 (43.4-46.9) N=7	32.5 (30.0-35.4) N=8
♂♂ (excluding obvious juveniles)	537.5 (300-700) N=4	272 (259-290) N=4	255 (241-268) N=4	46.5 (42.2-50.4) N=4	52.9 (52.1-53.9) N=3	38.1 (32.3-42.0) N=3

Antechinus cf. *bilarni*

Three (M15787-9) were collected from lines 1 (1♂) and 2 (2♀) in October 1976. These were collected from mixed low woodlands with a shrub layer with *Acacia* spp. prominent over *Triodia* sp. Hummock Grass on rugged sandstone.

Neither female was pregnant or lactating, although one had well developed teats. Because of the confusion concerning the taxonomy of Kimberley *Antechinus*, Mitchell Plateau specimens are not compared with others in the Western Australian Museum collections.

Basic body and skull measurements of these specimens are listed in **Table 1.**

Table 1: Measurements (in mm) of Mitchell Plateau *Antechinus* cf. *bilarni* and *Sminthopsis* cf. *douglasi.*

	Antechinus cf. *bilarni*			*Sminthopsis* cf. *douglasi*
WAM No	M15787	M15788	M15789	AMNH198645
Sex	♂	♀	♀	♂
Head and Body length	111.9	189.0	99.7	86.4
Tail length	94.1	—	83.3	103.2
Hindfoot length	15.8	13.9	14.6	18.7
Ear length	18.8	17.6	18.3	20.0
Weight (gm)	33.0	27.5	18.3	32.0
Skull				
Total length	29.4	—	—	29.2
Condylobasal length	27.5	—	—	26.4
Zygomatic breadth	17.1	16.2	—	16.6
Interorbital width	5.9	5.8	5.5	5.0
Palatal length	15.5	14.3	—	14.4
Incisive foramen length	2.5	2.7	2.7	2.5
Mastoid breadth	8.4	—	8.1	11.7
Bulla width	3.9	3.9	—	3.3
I^1 - M^4 length	13.8	12.9	13.1	14.3
C - M^4 length	10.5	9.9	9.9	10.9
Outside M^{1-1} breadth	11.8	7.1	7.0	7.5

Planigale maculata (Gould, 1851)

On 23 March 1977, a female (M15356) was trapped at line 58. This was Low Woodland A of *Eucalyptus miniata* and *E. tetrodonta* over Low Woodland B of *Livistona eastonii* on laterite. This species apparently occupies a range of habitat in northern Australia. Marlow (1958) reports it from sclerophyll and rainforest. McKenzie *et al.* (1977) captured a male in Drysdale River National Park in a "tall open shrubland of *Acacia* sp. over tussock grassland with occasional sandstone outcrops" and Davies (1960) states that they occupy *Pandanus* and *Melaleuca* fringe areas bordering the Adelaide River flood plain in the Northern Territory.

The Mitchell Plateau specimen, which is only the second from the Kimberley, is not pregnant and has a barely perceptible pouch area.

The stomach contained only several unidentified insect legs.

Body measurements:
1♀ W: 4.5 gm; H-V: 54 mm; T-V: 55 mm; HF: 9.4 mm; FA: 9.7 mm; E: 9.6 mm.

Sminthopsis virginiae (De Tarragon, 1847) **Red-cheeked Dunnart**

On 20 and 27 October 1976, 2 females (M15685-6) were collected from lines 35 and 27. These came from the only lines placed on black-soil plains. The vegetation was low woodland dominated by *Eucalyptus papuana* over *Melaleuca viridiflora* shrubs over cane grass.

The only other specimen of this species from Western Australia is a male (B1841) collected from beneath a log near Kalumburu, Kimberley, on 7 December 1965.

At Mitchell Plateau in October 1976, both females had pouch young: one with 5, the other with 8. (Crown-rump lengths ca. 12 to 14 mm).

Body measurements:
2♀♀ W: 19.5, 21.0 gm; H-V: 78, 78 mm; T-V: 94, 101 mm; HF: 18.0, 18.8 mm; FA: 18.2, 18.8 mm; E: 17.5, 17.5 mm.

Sminthopsis cf. *douglasi* Archer, 1979

An adult male (AMNH number 198645) was collected by W.H. Butler in riparian grassland at Camp Creek, near line 11, on 30 July 1973. This specimen appears to have greatest affinity with *S. douglasi* but may be an undescribed species. Technically *S. douglasi* must at present be considered *nomen nudum*. Body and skull measurements are listed in **Table 1.**

MURIDAE

Rattus tunneyi (Thomas, 1904) **Tunney's Rat**

In June 1976, 3 females (M14573-5) were trapped near line 48.

In October-November, 22 (M15620-37, M15861-3 and M15872) were trapped from the following lines: 31 (1♂), 32 (1♂, 1♀), 39 (1♂), 47 (1♂, 3♀), 48 (4♀), 49 (1♂), 50 (2♂, 5♀), 51 (1♂) and 54 (1♀).

In March 1977, 21 (M15311-31) were trapped from the following lines: 48 (2♀), 57 (5♀), 58 (5♂, 2♀), 59 (2♂, 3♀) and 60 (1♂, 1♀).

The June specimens were trapped in Open Low Woodland A over sedges next to a watercourse. In October-November 1976 this species was most readily captured in riparian Low Woodlands A or B over grasses or sedges with potable water nearby (17 of the 22 captures were in such situations). These situations are also characterised by deeper soils; this may be the important factor in determining their distribution in the Mitchell Plateau area, because Taylor and Horner (1973a) state that in the Cobourg Peninsula this species has extensive burrow systems and is most common in areas where the soil is both readily workable and supports good grass cover: "The opportunity to burrow must be vitally important to this rat". McKenzie *et al.* (1975, 1978) report similar burrow systems in the Kimberley. The remaining captures were in deciduous vine thicket and in Low Forest A on the edge of the plateau and in an ecotone between Low Open (eucalypt) Woodland and mangrove. Most were trapped from coastal sites. Only one was collected on other than volcanics — this was on black soil country at line 54.

In March 1977, although lines 48, 49 and 51 were placed in lowland vegetation fringing watercourses where the species had been collected the previous October, only 2 of the 21 specimens collected (from 576 trapnights) were from such areas. The remainder, from 640 trapnights, were collected from lines 57, 58, 59 and 60 sited on the lateritic plateau in Low Woodland A *over* Low Woodland B on laterite. These captures from the plateau in March 1977 were in marked contrast to the October-November 1976 situation when 574 trapnights, in very similar plateau habitat, resulted in no captures of this species. This suggests that distribution of *R. tunneyi* expands during the wet season to include higher ground. The Western Australian Museum has specimens of this species from many localities, indicating that *R. tunneyi* is widely distributed in the more northern parts of the Kimberley.

At Mitchell Plateau in June 1976 a lactating female with obvious uterine implantation scars and 2 juvenile females were collected. In October and November 1976, none of the females collected were pregnant or lactating and only 2 had uterine scars. Only one of the males captured had scrotal testes and these were of a size judged from Taylor and Horner (1973b) to be transitional in sperm production. There were no obvious juveniles in the collection. In March 1977, 5 of the 6 adult females collected were pregnant with from 8 to 10 foetuses; the crown to rump lengths ranged from 3.6 to 26.9 mm; enlargement of both mammae and teats suggested the other was nursing young. The 7 adult males collected at this time all had scrotal testes which ranged in length from 18.2 to 25.9 mm and were probably producing sperm. The remaining male and 7 females collected were juveniles ranging in weight from 29.0-55.0 gm.

Observations on *R. tunneyi* in Western Australian Museum collections from Boongaree I. in July 1973, Prince Regent River Reserve in August 1974, and Drysdale River National Park in August 1975, indicate considerable reproductive activity in July but perhaps diminished activity in August. For example at Boongaree I., 5 of the 9 females collected were pregnant and 3 were thought to be nursing young; all 9 males had scrotal testes (6 probably producing sperm, 2 transitional and 1 aspermic, based upon size). At Prince Regent River Reserve, of the 12 females collected none was pregnant and only 2 appeared to have bred. None of the 18 males collected had scrotal testes and their testes, on size, were aspermic. There were 6 obvious juveniles (HF 21.8-23.9 mm), including a virgin female in the collection. At Drysdale River National Park, 1 of the 11 females collected was pregnant and 5 had implantation scars but only slight mammary development. Only 3 of the 22 males collected were considered to be producing sperm, 4 had testes thought to be transitional and 18 were aspermic. There were 3 juveniles. Taylor and Horner (1973b), working largely from collections from the northern tip of Northern Territory indicate that this species breeds at least from May to August but state that paucity of collecting in other months prevents an estimate of breeding on a year-round basis. More recently, C.S.I.R.O. workers in Western Arnhem Land (Anon. 1974, p.10) indicate that *R. tunneyi* are marked seasonal breeders with the peak of breeding activity in the middle of the dry season.

A likely sequence of events for this species in the north and north-west Kimberley is that females give birth to young from the end of the wet season through the dry season, towards the end of which reproductive activity declines, population numbers fall and distributions retract to the moister areas surrounding the watercourses where the soil is sufficiently workable for them to burrow. Towards the end of the wet season and for several months following it, *R. tunneyi* is most active reproductively, population numbers are high and it is distributed most widely in the area.

The stomachs of 5 specimens examined had finely chewed fibrous plant material and occasional fragments of insect chitin.

It appears from the body measurements presented below that males are generally larger than females. This difference is probably even greater than is indicated by the tabulated measurements because some subadults may be included in the male sample.

Body measurements (mean and range):

	W (gm)	H-V (mm)	T-V (mm)	HF (mm)	FA (mm)	E (mm)
♀♀ (parous only)	79.2 (52.0-116.0) N=10	125 (113-142) N=9	116 (97-133) N=10	26.6 (25.0-28.6) N=11	23.2 (21.7-26.7) N=11	17.1 (15.4-18.5) N=5
♂♂ (excluding obvious juveniles)	97.8 (61.0-135.0) N=10	139 (113-163) N=11	124 (106-140) N=11	27.9 (26.5-29.9) N=11	24.3 (22.8-26.5) N=11	16.4 (14.0-18.8) N=15

Mesembriomys macrurus (Peters, 1876) **Golden-backed Tree-rat**

In June 1976 an adult female (M14567) was collected on a shed roof at Mining Camp surrounded by a Low Woodland A of *Eucalyptus miniata, E. tetrodonta* over a Low Woodland of *Livistona eastonii.*

In October-November 1976, 8 (M15791-97 & M15865) were collected from the following lines: 25 (1♂), 42 (1♂, 1♀), 45 (1♂). A female and its partially furred young were collected 0.25 km NW of line 30 in a Screwpine (*Pandanus* sp.) in riparian Open Low Woodland B on black soil plain. A female trapped in the roof-vent of a caravan at Walsh Point (near line 48) was also collected. Another female was collected at 2100 hrs 0.5 km W. Surveyors Pool from an *Owenia vernicosa* tree in Open Low Woodland B on rugged sandstone.

This species appears to occupy a diversity of habitats in the survey area. Specimens were obtained from *Eucalyptus/Owenia vernicosa* woodlands over spinifex on rugged sandstone and from *Eucalyptus/Pandanus* woodlands over grassland on black soil in the Surveyors Pool area; from the interphase between mangroves and deciduous vine thicket on basaltic scree at the Port Warrender headland site south of Walsh Point; from deciduous vine thicket and *Eucalyptus* woodlands on basaltic scree at Walsh Point, and from *Eucalyptus* and *Livistona eastonii* woodland on the lateritic plateau.

McKenzie *et al.* (1978) captured this species on islands in the Bonaparte Archipelago in hummock grassland on sandstone and in woodlands on volcanics. McKenzie *et al.* (1975) recorded 2 specimens in Prince Regent River Reserve from deciduous vine thicket on mixed sandstone and volcanic scree. Although Western Australian Museum collections have specimens from few coastal or near coastal localities (Kunmunya, Wotjulum, Kuri Bay, Prince Regent River Reserve, Wollaston I., Uwins I. and S.E. Osborne I.), this species is thought to be widespread in coastal parts of north-west and central northern Kimberley.

At Mitchell Plateau the female captured in June 1976 was neither pregnant nor nursing young. Of the 4 females collected between 18 October and 3 November 1976, 3 were not pregnant but had enlarged teats and were considered to be nursing young. The reproductive tract of the other was damaged when it was shot. One of these nursing females (M15792) had one teat more enlarged and flatter than its other teats. Its partially furred young was later recovered from a nest in the same *Pandanus* from which it was collected. The right uterine horn of this female contained 2 swellings ca. 5.8 mm in diameter. Histological preparation of the tract of this female revealed no pregnancies in the right horn and an apparently active corpus luteum in each ovary. The retention of corpora lutea for some time following birth is a feature of *Mesembriomys gouldii* (Crichton, 1969).

The Western Australian Museum has female specimens from the following dates: 9 June 1954 (M3019, Wotjulum Mission), 29 August 1974 (M12258A, Prince Regent River Reserve), 6-24 September 1955 (M3115-6 and M3118, Wotjulum Mission). These are skin specimens except for M12258A which is a spirit-entire specimen. The June female had small teats. The 3 females from September were subadult, one was only about half adult size. The August female had enlarged teats and 3 implantation scars on one uterine horn.

Combined, the observations from Mitchell Plateau and other Kimberley specimens suggest that this species gives birth to young at least from August through to October.

Stomachs of all 6 specimens examined contained vegetable matter and some insect chitin.

Body measurements (mean and range) (field recorded):

	N	W (gm)	H-V (mm)	T-V (mm)	HF (mm)	FA (mm)	E (mm)
♀♀ (parous only)	3	296 (238-330)	210 (199-229)	335 (318-360)	51.6 (49.7-54.0)	37.5 (34.0-40.0)	27.5 (25.7-29.3)
♂♂ (excluding obvious juveniles)	2	(207, 256)	(188, 201)	(291, 318)	(49.7, 51.6)	(35.3, 37.8)	(26.8, 29.1)

Conilurus penicillatus (Gould, 1842) **Brush-tailed Tree-Rat**

In June 1976, an adult male and an adult female (M14568-9) were shot at 1900 hrs near line 48.

Between October and November 1976, 9 (M15798-806) were collected from the following lines: 46 (1♂, 1♀), 48 (1♂, 3♀), 50 (1♀) and 51 (2♀).

On 16 March 1977, a female (M15354) was trapped at line 51.

All were captured on Carson Volcanics on the slopes of the Walsh Point headland. These were caught in Open Low Woodland A, Low Forest A over tall grass with *Eucalyptus miniata, E. tetrodonta, Bombax ceiba* var. *leiocarpum* and

Lysiphyllum cunninghamii prominent. Lines 48 and 50 were on a watercourse with occasional *Pandanus* sp. over tall grass.

The Western Australian Museum collections have a further 8 specimens from the Kimberley, and these are all from coastal localities.

The only other information on the habitat of this species in the Kimberley is from McKenzie *et al.* (1975) who record a specimen from low woodland on Carson Volcanics below the eastern face of Mt Trafalgar.

Calaby and Keith (1974) reported the species common in the Port Essington-Port Bremer area, N.T., in eucalypt forest and woodland and occasionally inside the margins of monsoon forest. It was also common in groves of *Casuarina equisetifolia* above the beaches. They report seeing them among the branches of trees and foraging on the ground. Several had stomachs containing green vegetable matter. Gilbert (in Gould 1851) reported seeing the species on beaches at night running along the edge of the surf, possibly to feed on animal matter brought in by waves. He describes its nest in *Casuarina* trees. In June 1976 one of us (KFK), whilst head torching at night, observed individuals foraging in tall grass near line 48 (2 collected).

At Mitchell Plateau the adult female from June 1976, had 3 uterine implantation scars, and both moderately enlarged teats and mammae; it had probably recently ceased nursing young. Three of the 5 females from late October-early November 1976, had uterine implantation scars (2, 2, and 4), but no obvious teat or mammary gland swelling; the other 2 females had no uterine scars or teat and mammary gland development. The female collected on 16 March 1977, had 3 moderate-sized foetuses with crown to rump lengths of ca. 18.7 mm. The male collected in June 1976 had moderately large scrotal/inguinal testes ca. 14.3 mm long; 2 of the 3 males collected in October had inguinal/abdominal testes ca. 9.4 and 13.5 mm long, respectively. The other October male had large scrotal testes ca. 16.7 mm long. The testes of none of these males are of a size considered by Taylor and Horner (1971) to be active. One of the females from the Western Australian Museum collections (M17279) was collected from Napier Broome Bay on 23 June 1910 with "3 young"—these young have not been located. The collections contain another 5 specimens which are juveniles or subadults collected from near South Alligator River, N.T., in May 1903 (specimen numbers 6145, 6749, 6148, 12817, 12816): 5 were very small with head to vent lengths between 72.0 and 95.8 mm; the other was subadult with head to vent length of 130.0 mm.

Combined, these observations from Mitchell Plateau and other Western Australian Museum collections suggest that *C. penicillatus* is a seasonal breeder, giving birth to young (2-4) at least from May through to June; by October reproductive activity has ceased. There are no females in our collections from December-February. Taylor and Horner (1971) report that in the Northern Territory this species is polyoestrous with a usual litter size of 2. They consider the breeding period extends "from May through, or even beyond August, if they are season breeders."

The stomachs of all 9 specimens examined were full of organic material. Seeds from an October specimen were most likely Papilionaceae and from a March specimen were possibly from a species of *Asclepias* (a climbing vine). Occasional pieces of insect chitin were found in most stomachs; these insects may have been ingested incidentally.

Body measurements (mean and range):

	W (gm)	H-V (mm)	T-V (mm)	HF (mm)	FA (mm)	E (mm)
♀♀ (parous only)	149 (112-168) N=4	151 (137-160) N=4	181 (164-194) N=3	41.7 (39.7-44.1) N=4	29.4 (28.5-30.0) N=4	26.3 (24.4-28.0) N=4
♂♂ (excluding obvious juveniles)	149 (115-132) N=4	149 (148, 151) N=2	190 (188, 192) N=2	43.3 (42.5, 44.1) N=2	29.5 (29.3, 29.7) N=2	25.7 (25.4, 26.0) N=2

Zyzomys argurus (Thomas, 1889) **Common Rock-Rat**

In June 1976, 3 males (M14577-9) were trapped near line 50.

In October-November 1976, 93 (M15537-15619, M15854-58, M15866-70) were trapped from the following lines: 1 (8♂, 1♀); 2 (2♂, 1♀); 3 (1♂); 11 (1♂); 22 (1♂); 23 (4♂, 2♀); 24 (2♂, 3♀); 25 (1♂); 26 (1♂, 2♀); 29 (1♂, 1♀); 33 (1♂, 1 unsexed); 34 (1♂); 37 (1♂, 3♀); 39 (4♂, 2♀); 40 (2♂); 41 (2♂); 42 (2♂); 43 (1♂); 44 (2♀); 45 (5♂, 1♀); 46 (3♂, 2♀); 47 (6♂, 2♀); 48 (3♀, 1 unsexed); 51 (2♂, 5♀); 52 (2♂, 5♀); 53 (1♂); 54 (1♀).

In March 1977, 5 were trapped from the following lines: 43 (1♀), 48 (1♂, 1♀, and 1 unsexed) and 56 (1♂). In addition 2 females were collected from the caravan site at Port Warrender, near line 48 (M15332-38).

Apart from its apparent absence from the lateritic plateau this species was widely distributed over the survey area in October-November 1976 and occupied a wide range of vegetation types. Some indications of habitat preference were obtained from captures per trapping effort in lines placed in sandstone *versus* volcanic substrate, hummock grass *versus* absence of hummock grass and near potable water (closer than 100 m) *versus* distant from potable water. There were significant relationships between captures and presence of hummock grass (X^2 = 16.7, $p < .005$) and sandstone substrate (X^2 = 4.09, $.05 > p > .01$). They were not caught significantly more frequently at lines near potable water than at lines placed distant from it.

The Western Australian Museum collection has specimens of this species from a number of widely distributed localities in the Kimberley, including Koolan Island, Prince Regent River Reserve, Kalumburu, Kununurra, Drysdale River National Park and Inglis Gap.

At Mitchell Plateau in June 1976, the 3 males collected were: an adult with scrotal testes 12 mm long, a subadult, and a juvenile. In October-November 1976, none of the females collected was pregnant although 18 of the 32 examined had implantation scars and 4 appeared to be lactating. None of the females were juveniles although the non-parous group were, on average, slightly smaller (31.3 gm, HF length 21.2

mm) than the parous group (37.2 gm, HF length 21.3 mm). In October-November 1976 only 4 of the 55 males were juveniles and weighed less than 20 gm; these had inguinal testes less than 4 mm long. Most of the other males had scrotal testes ranging in length from 3.4 to 11.7 mm with a strong correlation (R = 0.62, d.f. = 48, $p < .001$) existing between the weight of males and the length of their testes. The heavier males had longer testes. This suggests that many of these males were either subadult or recently adult. In March 1977, 2 of the females were pregnant; one with 3 foetuses which had crown to rump lengths of ca. 12.4 mm, the other with 1 foetus with crown to rump length of 28.3 mm. The other 2 females were lactating; one was nursing a partially furred young in a nest of shredded paper in a caravan cupboard (HV: = 45.8 mm, TV: = 34.8 mm), and was seen dragging this young behind it attached to a teat. Both males collected had scrotal testes, 10.5 and 12.0 mm long.

McKenzie *et al.* (1977) report that 10 of the 36 females collected from Drysdale River National Park in August 1975 were pregnant and that another 20 females had both enlarged uteri and mammae. McKenzie *et al.* (1978) report that 4 of the 12 females collected from islands of the north-west Kimberley coast in May 1972 were pregnant and that another 3 females had both enlarged uteri and mammae. Kitchener (1977) recorded a pregnant female at Lissadell Homestead, Ord River area, in late November 1976. The Western Australian Museum also has 2 pregnant females from Kalumburu collected by W.H. Butler on 22 December 1965 (B1908) and January 1966 (B2001). Both these females had 2 foetuses with crown to rump lengths ca. 25 mm. Three other females were collected at Kalumburu between these dates. These had slight uterine horn swellings but were not lactating (B1919, B1968 and B1986). Four males were collected during the same survey; they had scrotal testes which ranged in length from 9.6 to 12.0 mm.

Combined, the observations above generally agree with work by the C.S.I.R.O. Wildlife Division in western Arnhem Land which indicates that this species gives birth throughout the year (Anon., 1974), but with an indication that reproductive activity is considerably diminished towards the end of the dry season in the Kimberley.

Stomachs of 4 specimens examined contained finely masticated vegetable material.

Body measurements (mean and range):

	W (gm)	H-V (mm)	T-V (mm)	HF (mm)	FA (mm)	E (mm)
♀♀ (parous only)	39.4 (30-49) N=21	98 (83-117) N=21	117 (104-130) N=15	22 (20-23) N=22	18.4 (17.1-20.1) N=22	16.8 (15.2-18.2) N=20
♂♂ (excluding obvious juveniles)	34.9 (25.5-48.0) N=46	95 (81-106) N=44	107 (95-129) N=34	21.5 (18.9-23.6) N=48	17.9 (15.3-20.5) N=47	16.9 (15.1-19.5) N=40

At Mitchell Plateau, females appear to be slightly larger than males. In fact there is probably little difference in the size of the sexes because measurements of females were all from adults, (they had bred or were pregnant) whereas some subadults were almost certainly included in the male sample measured. Mitchell Plateau specimens appear to be slightly smaller than those from the Northern Territory recorded in Johnson (1964).

Zyzomys woodwardi (Thomas, 1909) **Woodward's Rock-Rat**

In October and November 1976, 29 (M15638-64 and M15859-60) were collected from the following lines: 24 (2♂, 3♀), 26 (3♀), 34 (2♀), 37 (2♂, 3♀, 1 unsexed, 1♀ released), 38 (1♀), 40 (2♂, 1♀), 41 (1♂, 2♀), 43 (1♂, 2♀), 44 (1♀), 45 (2♂) and 47 (2 released).

Between 13 and 16 March, 3 (M15339-41) were collected from lines 43 (1♀) and 56 (1♂, 1♀). At Mitchell Plateau this species appears to be about equally as common in, and largely confined to, both deciduous vine thickets (22 specimens from 2105 small mammal trapnights) and sandstone country excluding vine thickets (8 specimens from 1020 small mammal trapnights).

Most of the specimens from vine thickets were from the coastal volcanics, although 3 were trapped on sandstone scree slope near Surveyors Pool. The specimens from King Leopold Sandstones were from low woodlands with an upper stratum variously of *Owenia vernicosa, Eucalyptus* spp., *Terminalia* sp., *Planchonia australis, Xanthostemon paradoxus, Buchanania obovata* and *Vitex glabrata* over Scrub and Hummock Grass. The other 4 specimens were trapped in low woodlands with an upper stratum which included *Eucalyptus miniata, E. tetrodonta, Hakea arborescens, Grevillea* sp. and *Terminalia canescens* shrubs over tall grass on volcanics.

McKenzie *et al.* (1978) record this species in "all types of rugged King Leopold and Warton Sandstone country" as well as on Carson Volcanic scree covered with vine thickets.

We believe this species to have a patchy distribution in the more coastal areas of the Kimberley.

At Mitchell Plateau, 11 of the 17 females collected in October-November 1976 had uterine implantation scars and 4 of these had elongate teats, 5-9 mm long, and were thought to be lactating. The remaining 6 females had tiny teats and small uteri. They were considered subadult because they were generally smaller (mean HF length of 27.3 mm, range 25.8-28.3 mm) than the parous females (mean HF length 29.5 mm, range 27.7-30.7 mm). Eight of the 9 males had testes located intermediately between the inguinal and scrotal position; they varied in length between 4.1 and 7.7 mm. The other male was obviously subadult. It had abdominal testes 3.3. and 3.7 mm long. In March 1977, one of the females was pregnant with 2 foetuses with crown to rump lengths of ca. 7 mm; the other had enlarged teats and well developed mammae, but no obvious implantation scars. It is thought to have been recently nursing young (milk could not be expressed from teats).

Little is published concerning breeding of this species. McKenzie *et al.* (1978) record females with enlarged uteri and/or developed mammae on islands in the Bonaparte Archipelago in May, June, July and August; 3 females with foetuses in June and 5 with foetuses in July (2-3 foetuses). Examination of females in the

Western Australian Museum spirit collections showed that 1 of the 3 females collected at Prince Regent River Reserve in August 1974 had 2 foetuses and another was apparently lactating when collected. On 3 January 1966 a female (B1955), collected from Kalumburu, had 2 slight uterine swellings and both swollen mammae and teats; it is thought to have been nursing young.

It summary, recently parturient and/or pregnant females of this species were collected from the Kimberley in March, May, June, July, August, October and January—suggesting that this species gives birth to young throughout the year. The reproductive condition of the females from Mitchell Plateau in October-November 1976 suggests, however, that reproductive activity may be considerably reduced towards the end of the dry season.

The stomachs of 16 specimens examined contained fibrous vegetable material only.

Body measurements (mean and range):

	W (gm)	H-V (mm)	T-V (mm)	HF (mm)	FA (mm)	E (mm)
♀♀ (parous only)	152 (79-186) N=9	145 (133-155) N=11	124 (121-130) N=3	29.2 (27-31) N=10	25.1 (22.7-27.2) N=10	19.6 (17.2-21.4) N=9
♂♂ (excluding obvious juveniles)	128 (95-168) N=9	141 (121-166) N=8	120 (111-125) N=3	28.9 (28-31) N=8	21.2 (21.7-25.7) N=8	19.3 (18.3-20.3) N=7

Pseudomys forresti (Thomas, 1906) **Forrest's Mouse**

On 22 October 1976, a male (M15536) with scrotal testes 1.4 mm long, was trapped in line 13. This was a Low Woodland B over Dense Low Grass on volcanic slopes, with skeletal soils, about 400 m south of Camp Creek. The trees had been burnt recently and were suckering; *Eucalyptus* spp. and *Terminalia canescens* were dominant.

Little other information is available about this species' habitat in the Kimberley. Parker (1973) considers that in the Northern Territory it is seasonally fairly common on open black-soil plains of the Barkly Tableland but elsewhere is uncommon or rare. He lists its other habitats as open grassy plain, in the bases of *Triodia* clumps, and in the vicinity of soakages with stands of tall dense paperbark and eucalypt. It has been recorded from a few other localities in the Kimberley: near Kununurra, Kalumburu and Ningbing.

Body measurements:

1♂ W: 14.8 gm; H-V: NA; T-V: 50 mm; HF: 16.4 mm; FA: 14.3 mm; E: 14.0 mm.

Pseudomys delicatulus (Gould, 1842) **Little Native Mouse**

On 14 February 1973, L.A. Smith and R.E. Johnstone captured a male (M12182) with scrotal testes 3.0 mm long, by burning spinifex at Crystal Creek (near line 52).

This was an Open Low Scrub A of *Acacia orthocarpa* over spinifex on skeletal loamy soils with outcropping sandstone. Calaby and Keith (1974) report that this species is common in areas of sandy soil and is particularly plentiful on the inner side of the beach dunes covered with grasses, herbs, and low shrubs. It apparently does not occur in forested areas. The species is widely distributed throughout the Kimberley.

The Western Australian Museum has alcohol-stored specimens from the Kimberley collected on 29 April 1974 (Bohemia Downs Homestead, M12162-4); 3-6 June 1972 (Bonaparte Archipelago, M9318, M9320-4); 12 June 1974 (Beagle Bay Mission, M12165-6); 17 June 1965 (Inglis Gap, M7394); 2 July 1973 (Anjo Point, M10453); 24 July 1965 (Parry Creek, B1432-3); July 1967 (Kununurra, M7766); 3-16 August 1975 (Drysdale River National Park, M14271-M14295); 4 August 1963 (Erskine Range, M5703); 6 August 1960 (Beverley Springs Homestead, M4215); 6-21 August 1976 (Edgar Range, M15064-103); 20 December 1965 (Kalumburu, B1901-2); 1-16 January 1966 (Kalumburu, B1935-6, B1982-3).

The female collected in April beneath rubbish at Bohemia Downs Homestead was captured along with two young which it was either nursing or had recently nursed; its mammary glands and all its teats were enlarged (2 more so than the others). These young were fully furred with head to body length of 43 mm and hind foot length of 14.6 mm. Interestingly, the mother was pregnant with 4 moderately developed foetuses with crown to rump length of 11.8 mm. Another female from June was also pregnant (3 foetuses, uterine diameter ca. 2.7 mm) and had both moderately enlarged teats and mammae indicating that it had recently nursed young. Other pregnant females were collected in June at Beagle Bay Mission (3 and 4 foetuses, crown to rump lengths 2.7-3.4 mm), August at Erskine Range (1 foetus, crown to rump length of 2.5 mm) and Drysdale River National Park (4 and 2 foetuses—McKenzie *et al.* 1977). Females considered to be nursing young or to have recently been doing so were collected from these localities in June (2 specimens), August (5) and December (1); non parous females, probably subadults, were collected in June (2 specimens), July (1), August (5) and December (2). Only males collected in August and January had scrotal testes, but the location of testes appears unrelated to their size.

Combined, these observations suggest that in the Kimberley *P. delicatulus* is polyoestrous and gives birth to young throughout the year. Calaby and Keith (1974) report that *P. delicatulus* caught in July and August 1965 from Cobourg Peninsula, N.T., "were actively breeding and the majority of females were pregnant". Our observations indicating post partum oestrus in this species accord with those of Taylor and Horner (1970). They infer post partum oestrus in *P. delicatulus* on the basis that, among the females from Calaby and Keith's (1974) sample, which they observed to have implanted embryos, over half showed prominent development of teats suggestive of lactation.

Body measurements:

1♂ W: N.A.; H-V: 51.3 mm; T-V: 61.5 mm; HF: 15.9 mm; FA: 11.3 mm; E: 10.6 mm.

Pseudomys (sensu lato) sp.

On 24 October 1976 a female (M15535) was trapped at line 10. This was in a Low Woodland A with *Eucalyptus miniata* and *Planchonia australis* co-dominant with occasional *E. bigalerita* and *Livistona eastonii* over Dense Tall Grass on volcanics.

This species is intermediate in size between *Pseudomys delicatulus* and *P. forresti*. It probably represents a Kimberley population of a form obtained in Arnhem Land by J.H. Calaby which he proposes to describe. Another specimen attributable to this form was collected at Prince Regent River Reserve in valley eucalypt woodlands over *Sorghum* grasses (McKenzie *et al.* 1975), and a further 9 were collected in Drysdale River National Park, in woodlands of *Eucalyptus, Melaleuca*, and *Cochlospermum* over grassland on mudstone and sandstone (McKenzie *et al.* 1977).

Body measurements:

W: 6.5 gm; H-V: 54 mm; T-V: 61 mm; HF: 14.2 mm. Skull measurements: total length 20.3 mm; condylobasal length 17.2 mm; zygomatic breadth 10.3 mm; interorbital width 3.3 mm; palatal length 11.4 mm; nasal length 6.7 mm; mastoid breadth 8.2 mm; M^{1-3} length 4.2 mm.

Pseudomys nanus (Gould, 1858) **Western Chestnut Native Mouse**

In June 1976 a male (M14572) was caught by a dog near the Mining Camp in vegetation similar to line 18-20. Another male (M14576) was caught in tall grassland on the edge of the swamp in which line 21 was placed.

In October-November 1976, 36 (M15501-34, M15853 and M15873) were collected from the following lines: 1 (1♂, 1♀), 3 (5♂), 4 (4♂), 7 (1♀), 8 (6♂, 2♀, 1?), 9 (1♂, 2♀), 11 (1♀), 14 (2♂), 21 (4♂, 1?), 27 (2♂), and 36 (1♂, 1♀).

In March 1977, 3 (M15342-44) were caught from line 7 (2♂, 1♀).

In October-November 1976, approximately equal numbers, proportional to trapping effort, were collected from volcanic and lateritic situations. Relatively few (2 specimens) were from sandstone. By far the most (19 specimens) were from riparian Open Low Woodland A, with *Eucalyptus* spp., *Planchonia australis* and *Pandanus* sp. prominent over thickets over mixed Tall Grass. Examining only those lines from which the species was trapped, almost as many proportional to number of "small mammal" trapnights, were captured in the drier situations (9 specimens, 510 small mammal trapnights) as were captured in the moister situations, with potable water closer than 100 m (26 specimens, 1160 small mammal trapnights). The Dense Low Forest A of *Melaleuca leucadendron* on the dry swamp of line 21 is regarded as a moist situation. In June 1976 and March 1977 only 4 specimens were collected and these came from habitat similar to where they were collected in Spring 1976. These observations accord with those of McKenzie *et al.* (1977) who captured 43 *P. nanus* in Drysdale River National Park and record that the majority of specimens (not related to trapping effort) were collected from "grasslands . . . in and adjacent to watercourses. Upper storeys were often present as low open-forests and open woodlands . . . or as open forests".

P. nanus is widely distributed throughout the Kimberley.

The Western Australian Museum has *P. nanus* alcohol specimens collected on the following dates: 10 May 1968 (Wyndham, M8108); 26 May 1963 (Kununurra, M5521); 6 June 1965 (King Edward River, M6883); 6-22 August 1975 (Drysdale River National Park, M14227-M14270); 10-13 August 1974 (near Beverley Springs Homestead, M12310-11, M12306-7); 13-23 August 1974 (Prince Regent River Reserve, M12312-4, M12300-1, M12303-7, M12309); 23 August 1963 (Gogo Station, M5696); 3-8 December 1965 (Kalumburu, B1361, B1837-40); 14 January 1972 (Ord River, M11593-4) and 21 January 1973 (Mitchell River Station, M9853-4).

At Mitchell Plateau in June 1976, neither male collected had scrotal testes, one was subadult with testes 4.5 mm long and the other was an adult with testes 10.2 mm long. In October-November 1976, 5 of the 8 females collected had uterine implantation scars but no indication of mammae or teat development. Two had 'thread' uteri, were very small (20-20.5 gm) and were subadult. None was pregnant. Only 7 of the 26 males collected had scrotal testes and most of these testes did not appear to be fully developed (length 7.3 to 12.1 mm). The remaining males had testes from 4.0 to 8.9 mm long and several of these were probably subadult. In March 1977 the male had scrotal testes 14.7 mm long and the female was pregnant with 4 foetuses with crown to rump length of ca. 5.0 mm. Only 5 of the males collected elsewhere in the Kimberley in May, December and January had scrotal testes and of these only the 3 from January were fully enlarged (16.7, 17.6 and 18.4 mm long). Other males had inguinal testes. Of the 20 males examined from collections made in August all but 4 were obviously subadult; the adults had slightly darkened testes ca. 11 mm long. The May female was subadult. The June female had recently bred; it had uterine scars and was thought to have been lactating when collected. Of the 5 females collected in August, 3 were subadult, 1 had uterine implantation scars and slight mammary swelling and the other was probably either in oestrus or early pregnancy—with slight uterine swelling but little mammae development. In early December 1965, W.H. Butler collected 4 juveniles, each weighing about 10 gm, from a nest at Kalumburu. The female from January 1973 had 3 foetuses with crown to rump lengths of 5.7 mm. McKenzie *et al.* (1977) further report that at Drysdale River National Park in August 1975, 5 of the 19 female *P. nanus* they collected "had some uterine enlargement; one of these had seven uterine implantation scars and developed mammae."

In summary, pregnant or recently parturient females were observed in January, March, June, August and December. Subadults have been collected in May, June, August, October and November. This suggests that breeding takes place throughout the year in the Kimberley with the Mitchell Plateau observations providing evidence for reduced reproductive activity towards the end of the dry period.

Body measurements (mean and range):

	W (gm)	H-V (mm)	T-V (mm)	HF (mm)	FA (mm)	E (mm)
♀♀ (parous only)	33.6 (25.5-39.0) N=3	94 (80-104) N=3	94 (73-107) N=3	22.5 (22.2-22.7) N=4	17.0 (14.8-18.2) N=4	15.1 (14.6-15.7) N=3
♂♂ (excluding obvious juveniles)	33.9 (25.5-46.5) N=16	96 (83-115) N=14	101 (85-119) N=12	23.3 (21.8-24.4) N=16	17.9 (16.8-18.7) N=15	15.0 (13.5-16.4) N=12

Melomys cervinipes (Gould, 1852) **Fawn-footed Melomys**

In October and November 1976, 21 (M15665-85) were collected from the following lines: 30 (3♂, 3♀), 31 (3♂, 3♀), 43 (3♂, 1♀), 44 (1♂, 1♀) and 50 (2♂). Trapping information is unavailable for the male M15674. Specimens were released from the following lines: 31 (3 ♂, weights: 79.0, 87.0 and 114.0 gm; 2 ♀, weights: 71.0 and 90.0 gm) and 52 (one, sex unknown, escaped).

On 11 March 1977 an adult male (M15345) was shot from a twig in deciduous vine thicket near line 43.

On 15 March 1977 an adult male (M15346) was trapped at line 55.

This species has a clear preference for the deciduous vine thickets at the edge of the lateritic plateau and beneath the volcanic scree on the coast; 24 of the 26 captures were from these habitats. The remaining 2 captures were along watercourses in Open Low Woodland A of *Eucalyptus* sp. and *Dolichandrone* sp. over *Pandanus* and sedges on volcanics, and in a Forest of *Melaleuca leucadendron* over *Pandanus* and Tall Grass.

At Mitchell Plateau in October and November 1976, 5 of the 8 females collected were pregnant, each with 2 foetuses. These ranged from very recent to near-term pregnancies (foetus crown-rump lengths of 3.6 to 18.1 mm). Of the other 3 females collected, 1 appeared to have recently ceased nursing young; 1 was a non-parous adult, and 1 a subadult. All 13 males collected in October-November appeared to be adults. They had scrotal testes which ranged in length from 16.3 to 22.0 mm. The males collected in March 1977 had scrotal testes 15.5 and 18.2 mm long.

The stomachs of 4 specimens examined contained chewed plant material only.

From the measurements tabulated below there appears to be little difference between the general body size of males and females.

Body measurements (mean and range):

	W (gm)	H-V (mm)	T-V (mm)	HF (mm)	FA (mm)	E (mm)
♀♀ (parous only)	74 (53-85) N=5	125 (111-135) N=5	157 (149-166) N=5	28.2 (27.8-29.2) N=5	23.4 (23.2-24.5) N=5	17.5 (17.4-18.5) N=5
♂♂ (excluding obvious juveniles)	88 (74-110) N=12	126 (118-138) N=10	153 (149-163) N=11	28.8 (27.7-30.3) N=13	23.7 (22.6-25.3) N=13	17.6 (16.4-18.5) N=13

VESPERTILIONIDAE

Nyctophilus arnhemensis Johnson, 1959 **Arnhem Land Long-eared Bat**

Between 2-4 November 1976, 3 males and 1 female (M15776-9) were mist-netted at about 2200 hrs among mangroves at Walsh Point (opposite lines 42 and 49).

On 14 March 1977 a male (M15351) was mist-netted at 1830 hrs among mangroves at Walsh Point (opposite line 48).

The capture of this species amongst and under the dense foliage of mangroves near Mitchell Plateau extends its described habitat. Johnson (1964) observed this bat close about foliage in more densely forested areas, particularly in tunnel-like passages in monsoonal forests. He reports that its daytime refuge sites are often beneath the bark of paperbark trees (*Melaleuca leucadendron*). It has also been collected along the Drysdale River, Kimberley, in a "fringing" formation of *Melaleuca argentea, M. leucadendron* and *Pandanus* (McKenzie *et al.* 1977) and over dams and lagoons in the Ord River area and Northern Territory (Kitchener 1978, Calaby and Keith 1974 respectively). In addition to the above specimens the other Western Australian Museum specimens are from Drysdale River National Park.

At Mitchell Plateau the female collected in November appears to be recently pregnant with one horn slightly swollen to 3.7 mm diameter. Kitchener (1978) reports that two females collected in the Ord River Valley in November 1976 had recently given birth to young.

Mitchell Plateau specimens appear to have shorter bodies but longer tails, forearms and ears than those from Northern Territory recorded by Calaby and Keith (1974) and Johnson (1964).

Body measurements (mean and range):

	N	W (gm)	H-V (mm)	T-V (mm)	Ti (mm)	FA (mm)	E (mm)
♀♀ (parous)	1	6.0	48	45	20.2	40.9	22.5
♂♂ (excluding obvious juveniles)	3	6.0 (4.5-7.5)	46 (44-47)	40 (39-43)	19.2 (18.7-19.7)	38.0 (37.8-38.0)	21.6 (21.0-27.6)

Nyctophilus walkeri Thomas, 1892 **Walker's Long-eared Bat**

On 23 June 1976, a female (M14580) was shot over a waterhole fringed with *Pandanus aquaticus* on Camp Creek near line 7. These are the third and fourth specimens of this species collected in Western Australia. The other 2 specimens were males; they were collected over a sandbank in fringing formation of *Melaleuca* and *Pandanus* in Drysdale River National Park along the Drysdale River on Pentecost Sandstone (McKenzie *et al.* 1977).

On 19 October 1976, a female (M15768) was mist-netted over a waterhole on Camp Creek near line 5. Both these waterholes were fringed by *Pandanus aquaticus* and *Melaleuca leucadendron* on volcanics.

At Mitchell Plateau, the June specimen was neither pregnant nor lactating. The October specimen had a foetus in each horn; these had crown to rump lengths of ca. 12.0 mm.

The stomach contained Coleoptera, Hymenoptera and Hemiptera.

Body measurements (field recorded):
1♀ W: 4.5 gm; T-V: 35 mm; H-F: 7 mm; FA: 35.1 mm; E: 13.0 mm.

Miniopterus schreibersii (Kuhl, 1819) **Bent-wing Bat**

Between 18-20 October 1976, 1 male and 3 females (M15772-5) were mist-netted over Camp Creek near lines 1 and 7. The bank of the creek at both these sites is fringed with *Melaleuca leucadendron* and *Pandanus*; line 1 is on sandstone and line 7 on volcanics.

On 25 January 1973 a male (M10315) was collected from Crystal Creek by L.A. Smith and R.E. Johnstone.

In the Kimberley, this bat has been collected from large colonies in buildings and caves in the dry season when it is thought to be relatively inactive. It is infrequently collected at other times of the year. Parker (1973) records it as locally common in the Northern Territory, in deep caves, mine shafts and abandoned buildings.

At Mitchell Plateau in October, 2 of the 3 females collected were pregnant; each had a single foetus with crown to rump length of 11.8 and 13.9 mm respectively. The other female was not pregnant. None was lactating.

The Western Australian Museum has further Kimberley females from 19 June-4 July 1960 (Kalumburu, 21 specimens within M4847-76), 28 June 1974 (Lawford Ranges, M7052A-J), 19 July 1965 (Yeeda, M7441), 27 July and 2 August 1965 (Parry Creek, M7440, B1451), 9 August 1965 (Ningbing, M7442), 18 August 1974 (Prince Regent River Reserve, M12241) and 16 September 1975 (Drysdale River National Park, M14085, -87, -90 and -93). None of these was lactating and only 3 from Lawford Ranges on 28 June 1974 showed a slight swelling in the right horn.

Combined, these observations suggest that young are born about December. In New South Wales, Dwyer (1963) concluded that this species copulates in autumn, probably carrying an embryo in diapause condition through winter while in partial hibernation, and gives birth to young in mid-summer.

Stomachs contained Lepidoptera and Orthoptera.

Body measurements:

	N	W (gm)	H-V (mm)	T-V (mm)	Ti (mm)	FA (mm)	E (mm)
♀♀ (parous only)	2	8.3, 9.5	48, 55	46, 48	18.2, 18.3	43.5, 43.6	10.0, 10.8

Eptesicus pumilus (Gray, 1841) **Little Bat**

Between 19 October and 6 November 1976 5♂ and 7♀ (M15754-61 and M15876-9) were collected. During the day these were mist-netted or shot in the following sites: a small sandstone cave near line 1; a large system of caves in rugged sandstone country 1 km W. Surveyors Pool; a small cave in an area surrounded by vine thicket at the base of volcanic cliffs near line 37; and in a crevice further up the same cliff. During early evening they were shot over Surveyors Pool, and at a Crystal Creek pool near line 52.

On 23 January 1973 1♂ and 5♀ (M9845-50) were collected by L.A. Smith and R.E. Johnstone from a sandstone crevice at Crystal Creek near line 52.

This is a common bat in the Kimberley and the Western Australian Museum has specimens from numerous localities throughout the region. It has been recorded in Fairy Martin nests and in both sandstone and volcanic crevices and caves of all sizes.

At Mitchell Plateau in October-November 1976, the 7 females collected were pregnant; 5 had twins and 2 had a single foetus. The foetuses' crown to rump lengths ranged in size from 8.7 to 14.0 mm. The small group collected at Crystal Creek in January consisted of a lactating female and 4 subadults. Observations on breeding colonies of this bat in the Ord River area (Kitchener 1978) and at Tennant Creek, N.T. (Maddock and McLeod 1976) indicate that this species gives birth to young throughout the year.

Two specimens had Lepidoptera, Hemiptera? and Trichoptera? in their stomachs; the others had empty stomachs.

Mitchell Plateau specimens are generally smaller than those from the Northern Territory reported in Johnson (1964).

Body measurements (mean and range):

	W (gm)	H-V (mm)	T-V (mm)	Ti (mm)	FA (mm)	E (mm)
♀♀ (adults)	3.6 (3.3-4.0) N = 4	35 (31-40) N = 7	32 (28-34) N = 7	11.7 (11.2-12.7) N = 7	30.1 (29.3-31.0) N = 7	10.1 (9.5-10.8) N = 7
♂♂ (adults)	3.0 (2.5-3.3) N = 4	34 (32-35) N = 5	30 (26-32) N = 5	11.5 (10.6-12.2) N = 5	29.4 (28.5-30.5) N = 5	9.8 (9.0-10.4) N = 5

Eptesicus douglasi Kitchener, 1976 **Douglas' Little Bat**

On 22 October 1976 a male (M15770) was shot at 1810 hrs in a watercourse 0.5 km W. Surveyors Pool. A female (M15771) was mist-netted on 4 November over Surveyors Pool. This bat has a wide distribution in the north-west Kimberley.

At Mitchell Plateau in November the female was pregnant with a moderately large foetus with crown to rump length of 10.7 mm. Kitchener (1976) reports that females from August, June and October were neither pregnant nor lactating, which suggests that the period of births begins towards the end of the dry season.

Body measurements:

1♂ W: 5.0 gm; H-V: 39 mm; T-V: 34 mm; Ti: 13.7 mm; FA: 36.2 mm; E: 12.5 mm.

Myotis adversus (Horsfield, 1924) **Large-footed Myotis**

In October-November 1976 a male (M15766) was shot at 2105 hrs over Surveyors Pool. Four males (M15762-5) were shot at a Crystal Creek pool near line 52 at 1200 hrs: 3 of these were clustered together in a very slight depression in a sandstone wall near a small waterfall in Crystal Creek, the other male was collected ca. 5 m away, on a rock overhang just to the side of a waterfall.

The only other roosting site recorded for this bat in the Kimberley was also under an overhang in rugged sandstone country in the Drysdale River National Park, where McKenzie *et al.* (1977) collected 2 males and 15 females from 3 Fairy Martin nests. Despite extensive collecting in the Kimberley over the last 5 years it has not been observed in caves, which McKean and Hall (1965) consider to be its normal roosting sites in southern Australia. They also report it roosting under buildings in north-eastern Australia.

According to Dwyer (1970) this species generally forages over water. Although infrequently collected, it is probably widely distributed along watercourses of coastal Kimberley.

The Western Australian Museum has a collection of 15 females from Drysdale River National Park from August 1975. These had slightly enlarged uterine horns and small teats and were probably in oestrus or early pregnancy.

The stomachs contained Blattodea, Hemiptera (probably Gerridae), Lepidoptera, Diptera and Coleoptera.

Mitchell Plateau specimens appear to be generally smaller than those from Northern Territory recorded in Calaby and Keith (1974).

Body measurements (mean and range):

	W (gm)	H-V (mm)	T-V (mm)	Ti (mm)	FA (mm)	E (mm)
♂♂ (adults)	8.1 (7.0-9.5) N=5	45 (44-47) N=3	35 (34-36) N=3	17.4 (16.7-18.3) N=3	36.9 (36.7-37.2) N=3	13.7 (13.1-14.0) N=3

HIPPOSIDERIDAE

Hipposideros stenotis Thomas, 1913 **Lesser Warty-nosed Horshoe Bat**

On 25 January 1973, a female with attached partially furred young (forearm 12.5 mm) (M9851-2) was collected by L.A. Smith and R.E. Johnstone from a sandstone crevice at Crystal Creek.

On 17 February 1973, 1 female (M10711) was taken from beneath a sandstone overhang at Surveyors Pool by the above two collectors.

On 25 October 1976, a male (M15875) was shot over Surveyors Pool in sandstone country.

On 6 November 1976, 2 females (3172 and 3173*) were collected from a large deep sandstone cave 1 km W. Surveyors Pool.

At Mitchell Plateau in November 1976, one female was apparently lactating and had one uterine horn, slightly swollen to 3.2 mm diameter, which appeared to have recently involuted; the other female was neither pregnant nor lactating. The January female had recently given birth. The one collected in February appeared to be lactating but was not pregnant. The Western Australian Museum collections have 2 females (M10471-2) collected from Boongaree I., north-west Kimberley, on 5 July 1973; one of these was in early pregnancy with one uterine horn slightly swollen to 3.1 mm. The other was not pregnant and was not lactating when collected.

*Field numbers: Field Museum of Natural History, Chicago.

Combined, these few observations suggest that this species is in early pregnancy in July. The birth season of the single young commences about October and continues at least to the end of January, but more observations are required to confirm this suggestion.

The stomach of the 3 collected in October contained mostly Lepidoptera and Coleoptera (Scarabaeidae) with some Hymenoptera.

Body measurements (mean and range):

	W (gm)	H-V (mm)	T-V (mm)	Ti (mm)	FA (mm)	E (mm)
♀♀ (adult)	—	42 (41-43) N=4	26 (25-28) N=4	21.2 (20.4-22.4) N=3	44.9 (44.8-45.0) N=3	20.2 (19.0-20.9) N=4
♂ (adult)	5.2	39	24	20.3	45.6	19.1

Hipposideros ater Templeton, 1848 **Dusky Horseshoe Bat**

An adult male and adult female (M10308-9) were collected at Crystal Creek near Crystal Head on 25 January 1973. The female was not active reproductively.

The Western Australian Museum has specimens from only 3 other Kimberley localities: Augustus I., Byam Martin I. and the Ord River area.

Body measurements (mm):

1♀ and 1♂, respectively: W (gm); N.A.; H-V: 32, 34; T-V: 25, 21; Ti: 15.4, 14.5; FA: 35.5, 34.8; E: 16.9, 16.7.

Rhinonicteris aurantius (Gray, 1845) **Orange Horseshoe Bat**

On 2 November 1976, a male (M15767) was mist-netted at 1900 hrs over a pool in Crystal Creek, near line 53, fringed with *Pandanus, Melaleuca leucadendron* and *Acacia orthocarpa.*

The Western Australian Museum has few specimens of this species (near Kununurra, M6076 and M10023 and Koolan Island, M10054); it is, however, locally very common in the Kimberley. L.E. Keller observed a large colony of 200-300 *R. aurantius* in a humid inner chamber of a limestone cave near Geikie Gorge, south-western Kimberley. Parker (1973) reports that in the Northern Territory it is widespread and common near the coast and that its distribution may change seasonally.

Body measurements:

1♂ W: 4.0 gm; H-V: 50 mm; T-V: 26 mm; Ti: 19.7 mm; FA: 48.7 mm; E: 12.4 mm.

MOLOSSIDAE

Tadarida jobensis (Miller, 1902) **Northern Mastiff Bat**

On 3 November 1976 an adult female was mist-netted near line 53 over a small pool fringed by *Melaleuca leucadendron* and *Pandanus* sp. This specimen was neither pregnant nor lactating.

The Western Australian Museum has collections of this species from 2 other Kimberley areas both considerably drier than Mitchell Plateau: Derby (M7523-30 and M7485-6) and Drysdale River National Park (M14024-5, M14026 and 8, M14029-30 and M14032-3).

The stomachs contained Coleoptera, Lepidoptera, and Hymenoptera (possibly sawfly adults).

Body measurements:

1♀ W: 21.5 gm; H-V: 67 mm; T-V: 40 mm; Ti: 17.0 mm; FA: 49.1 mm; E: 20.4 mm.

EMBALLONURIDAE

Taphozous georgianus Thomas, 1915 — **Common Sheath-tailed Bat**

Between 19 October and 4 November 1976, 11 males and 23 females (M15686-719) were collected. During the day, some were mist-netted in a small sandstone cave near line 1. Others were shot in a small cave in an area surrounded by vine thicket at the base of volcanic cliffs near line 37, or mist-netted in a crevice further up the same cliff. During early evening some were shot over or near Surveyors Pool in rugged sandstone country, over a pool in laterite near Mining Camp or near line 7 in open eucalypt woodland on laterite. Although this species was not collected during the June 1976 or March 1977 trips to the Plateau it is a common and widespread species in the Kimberley.

At Mitchell Plateau in October-November 1976, 20 of the 23 females collected were pregnant (19 in the right uterine horn only and 1 in the left uterine horn only). None was lactating. Most of these were in the earlier stages of pregnancy and only 2 had foetuses with crown to rump lengths greater than 15 mm (18.1 mm and 20.5 mm respectively). At term, the crown to rump lengths are in excess of 33 mm (Kitchener 1973). There appeared to be a reasonably close synchronisation of fertilisation between these individuals. Of the 3 females which were not pregnant, 2 appeared subadult.

The stomachs of 4 specimens contained Coleoptera (Scarabaeidae, Rutelinae, and numerous other families), Lepidoptera, Heteroptera, Homoptera, Orthoptera and Diptera.

Specimens from Mitchell Plateau appear to have longer tails than those from Northern Territory reported in Johnson (1964).

Body measurements (mean and range):

	W (gm)	H-V (mm)	T-V (mm)	Ti (mm)	FA (mm)	E (mm)
♀♀ (adults)	18.1 (14.5-21.5) N=22	73 (68-78) N=23	28 (25-33) N=23	27.1 (25.9-28.3) N=23	67.0 (63.3-70.4) N=22	20.3 (17.9-21.8) N=23
♂♂ (adults)	17.1 (14.0-19.0) N=11	73 (68-77) N=11	28 (26-30) N=11	27.0 (25.3-27.8) N=10	66.9 (63.5-68.7) N=10	20.4 (19.5-21.7) N=11

PTEROPODIDAE

Pteropus alecto Temminck, 1837 **Black Flying Fox**

In October-November 1976, 1 female and 5 males (M15780-4 and M15864) were shot over a pool 1 km E. Mining Camp; at Surveyors Pool; at Camp Creek near line 1; over mangroves at Walsh Point near line 48; and over the beach near line 41.

In March 1977 a female (M15352) was mist-netted in mangroves at Walsh Point near line 48; no other individuals were seen during the March 1977 survey.

In October-November the species was common in groups of a few to several hundred individuals in *Melaleuca leucadendron* along watercourses in sandstone volcanic and laterite and black-soil plain country. Others were recorded in coastal situations, over mangroves and beaches. McKenzie *et al.* (1975, 1977) recorded this species in similar congregations in a similar variety of habitats during the north and north-western Kimberley dry season. They recorded a camp of >30,000 individuals in paperbarks along the Drysdale River in August 1975. The dramatic decline in numbers of this species in the survey area in March 1977 compared to October and November 1976 accords with Nelson's (1965) observations on *P. alecto (gouldii)* from Queensland. He considers this species congregates in large camps from early until late summer during which time young are born and raised. By April the numbers of individuals in the colonies falls dramatically as individuals disperse; immature animals form winter camps.

Neither female was pregnant nor had enlarged teats. Ratcliffe (1931) states that this species, unlike other Australian flying foxes, does not exhibit any obvious seasonal migrations although they do move locally.

Body measurements (mean and range):

	N	W (gm)	H-V (mm)	FA (mm)	Ti (mm)	E
♀♀ (adult)	1	440.0	200	—	90	34
♂♂ (adult)	3	210 (200-230)	205 (186-230)	172 (162-178)	85 (81-88)	33 (29-35)

Macroglossus minimus (Geoffroy, 1810) **Northern Blossom Bat**

As a result of Start's (1974) appraisal of the genus *Macroglossus*, we consider that the Mitchell Plateau specimens should be attributed to *Macroglossus minimus* of which *M. lagochilus* is a synonym.

Between 18 October and 4 November 1976, 34 (M15720-53, M15880) were mist-netted, between 1900-2300 hrs, from the following locations: over Camp Creek near lines 1 and 7 (4♂, 10♀); Surveyors Pool (3♂, 2♀); near Surveyors Pool (2♂, 4♀); and at Walsh Point; the headland site near line 47 and the western arm of the Lawley River Delta.

In March 1977 a male (M15353) was mist-netted in mangroves opposite line 48 in *Avicennia marina* and *Lumnitzera racemosa.*

At Mitchell Plateau this bat was present in the coastal mangroves (of *Aegiceras corniculatum, Ceriops tagal* and *Avicennia marina*), and along the more inland watercourses in sandstone, volcanic and black-soil plain country where they were seen feeding in large numbers in heavily flowering *Melaleuca leucadendron.*

This species, although rare in Western Australian Museum collections, is probably common along the Kimberley coastline. Parker (1973) reports *M. lagochilus* (≡ *minimus*) as locally abundant in the Northern Territory with an apparent peak in numbers in spring and early summer. McKean (1972) reports that in Papua New Guinea this species occurs in a range of forested areas.

Histological preparation of the reproductive tract of 4 female *M. minimus* from Mitchell Plateau, coupled with examination of reproductive organs *in situ* of other females from Mitchell Plateau collected in October-November, showed that their reproductive condition ranged from pro-oestrus (M15752), carrying recently implanted blastocysts (M15749) or late stage foetuses (M15738), recently parturient with a large corpus luteum in the right ovary, and less recently parturient with a large corpus luteum in the left, which was invaded by stromal cells and blood vessels. In addition we have examined a further specimen (CM4674) collected on 14 July 1971 at Mt Brockman Range, N.T., which has a moderate-size foetus. These observations indicate considerable reproductive activity during the dry season in the Kimberley. The histological evidence from two recently parturient and lactating females from Mitchell Plateau in October does not suggest post-partum oestrus in this species; neither of these females had ovaries with enlarged or vesicular follicles.

M. minimus is a nectar- and pollen-feeder and the identity of the pollen found in the stomachs of those collected at Mitchell Plateau is being examined by A.N. Start.

Body measurements (mean and range):

	W (gm)	H-V (mm)	Ti (mm)	FA (mm)	E (mm)
♀♀ (adults)	12.5 (8.0-19.5) N = 17	59 (49-67) N = 14	17.6 (15.9-19.2) N = 14	40.0 (38.1-42.6) N = 14	14.3 (12.9-15.0) N = 14
♂♂ (adults)	12.1 (10.3-14.5) N = 13	59 (57-65) N = 9	17.4 (14.7-18.8) N = 9	40.4 (39.5-42.6) N = 9	13.8 (12.7-16.0) N = 9

CANIDAE

Canis familiaris Linnaeus, 1758 **Dingo**

In June 1976 dingoes were seen at Walsh Point and near the Mining Camp site.

On 18 October, 4 dingo pups were sighted and 2 were collected at 1900 hrs near line 6 in low eucalypt woodland on volcanics. They were about 6 weeks old. During October and November dingoes were sighted in sandstone country south of Surveyors Pool; on the laterite plateau ca. 1 km S. Surveyors Pool; on volcanic slopes near Port Warrender; near Camp Creek and on black-soil country near Surveyors Pool. An emaciated dingo had trapped itself at the base of the Surveyors

Pool waterfall and an adult dingo was seen swimming in Camp Creek attempting to drag a dead and bloated Sandy Wallaby (*M. agilis*) from the water.

In March 1977 dingoes were sighted on the laterite plateau both near the airstrip and ca. 29 km from Port Warrender. A subadult was also sighted on volcanics near Camp Creek.

L.A. Smith and R.E. Johnstone saw several dingoes on the lateritic plateau near the Walsh Point road junction during their January-February 1973 expedition to the area. They shot one of these dingoes and found a bandicoot (*Isoodon macrourus*) in its stomach. In mid-January 1973, R.E. Johnstone captured a 5-8 week old dingo pup near line 7 in low open woodland on volcanics.

Most of the dingoes seen were in poor condition. The body colour varied from yellow-brown to grey-black. The dingo is widespread in the Kimberley and appears to be common throughout the survey area at Mitchell Plateau. Observation of dingoes eating bandicoots and a dead Sandy Wallaby at Mitchell Plateau are consistent with Whitehouse's (1977) statement that the dingo is an opportunistic feeder, capable of killing a wide range of prey species but not averse to feeding on carrion. He considers that in the rangeland areas of Western Australia, Red Kangaroo (*Megaleia rufa*) and Euro (*Macropus robustus*) form the largest part of their diet.

In the Kimberley, female dingoes with litters ranging in age from 5-8 weeks, have been seen at Mitchell Plateau in mid-January and mid-October, at Drysdale River National Park in August (McKenzie *et al.* 1977) and at the Ord River in mid-October (Kitchener 1978). Newsome *et al.* (1973) state that dingoes have a well defined seasonal reproductive cycle and breed once a year. In their studies, conducted in central Australia and south-eastern Australia, dingoes whelp in winter months, but they comment that latitude may effect the timing of breeding.

TACHYGLOSSIDAE

Tachyglossus aculeatus (Shaw, 1792) **Echidna**

In October 1976 numerous droppings, unquestionably from this species, were observed at lines 1 and 25. Both these lines were on sandstone; one vegetated with hummock grass, the other in a Woodland of *Terminalia* sp. over *Acacia* sp. and *Bossiaea bossiaeoides* shrubs over grassland.

This species has been collected or its sign observed from several widely separated localities in the Kimberley: Kalumburu, M7647; Black Elvire Creek, M7414; Ord River Valley, M2871; Prince Regent River Reserve, McKenzie *et al.* (1975); and Drysdale River National Park, McKenzie *et al.* (1977). Nowhere has it been observed in large numbers. Johnson (1964) states that this species is uniformly but sparsely distributed over all of Arnhem Land.

FELIDAE

Felis catus Linnaeus, 1758 **Feral Cat**

In October 1976 one animal was seen near line 13 in Low Woodland of *Eucalyptus* spp. and *Terminalia canescens* over Dense Low Grass.

In March 1977, 1 was seen on the edge of the airstrip in Low Woodland A of *Eucalyptus miniata* and *E. tetrodonta* over *Livistona eastonii* on laterite.

Feral cats are now well established in the Kimberley mainland. McKenzie (1978) does not, however, record them from any of the immediate offshore islands.

DISCUSSION

(a) Mitchell Plateau mammal assemblage

The Mitchell Plateau area has the largest number of species (39) collected from any area of similar extent in Western Australia. The bats *Macroderma gigas* (Dobson, 1880) and *Taphozous flaviventris* Peters, 1867 were sighted, but not collected, at Crystal Creek in November 1978 by W.H. Butler and at Camp Creek in October 1976 by L.E. Keller, respectively.

Collections from other north-western Kimberley coastal localities with a similar climate, vegetation and geology as Mitchell Plateau, viz. Prince Regent River Reserve (McKenzie *et al.* 1975), Kalumburu (WAM collections) and Kimberley Islands (McKenzie *et al.* 1978), indicate that a further 9 species may occur in the Mitchell Plateau area. These are: *Chalinolobus gouldii* (Gray, 1841); *C. nigrogriseus* (Gould, 1856); *Nycticeius greyii* (Gray, 1844); *Pteropus scapulatus* Peters, 1862—recorded from Mitchell River Station, 40 km south of the Mitchell Plateau; *Petropseudes dahli* (Collett, 1895); *Isoodon auratus* (Ramsay, 1887); *Mesembriomys gouldii* (Gray, 1843); *Hydromys chrysogaster* Geoffroy, 1804 and *Peradorcas concinna* (Gould, 1842). It is possible then, that of the 64 mammal species reported from the Kimberley, as many as 50 species may occur in the Mitchell Plateau area.

Comparison between the Mitchell Plateau collections and those from Prince Regent River Reserve (McKenzie *et al.* 1975), Drysdale River National Park, Kalumburu, and Inglis Gap (all cited in McKenzie *et al.* 1977), Kimberley Islands (McKenzie *et al.* 1978) and the Ord River Area (Kitchener 1978), indicates that the Mitchell Plateau area is not lacking any major elements of the Kimberley fauna.

Several species collected at Mitchell Plateau are of particular note in that they are collected from few localities in the Kimberley. These are the rock-wallaby, *Petrogale burbidgei*; the dasyurids *Sminthopsis virginiae, S.* cf. *douglasi* and *Planigale maculata*; the possums *Trichosurus arnhemensis* and *Wyulda squamicaudata*; the rodents *Mesembriomys macrurus, Conilurus penicillatus, Pseudomys forresti* and *Pseudomys* sp.; the bats *Nyctophilus arnhemensis, N. walkeri, Hipposideros stenotis, H. ater, Rhinonicteris aurantius* and *Macroglossus minimus.*

(b) Mitchell Plateau mammal habitats

The mammal richness of the Mitchell Plateau area (and north-western Kimberley generally) compared with other parts of the Kimberley is almost certainly the result of the wide range of physiography in the area coupled with a protracted wet season. This has resulted in the concentration of a wide diversity of habitats representative of many parts of the Kimberley, with, in addition, deciduous vine thickets which are rare outside the area. Another important factor is that the north-western parts of the Kimberley have not had extensive grazing by domestic stock—the presence of which Kitchener (1978) considers may have contributed to the low number of mammal species recorded as extant in the Ord River area.

We recognised 19 broad habitat types in the Mitchell Plateau area on the basis of groupings which considered substrate (sandstones, volcanics, laterites, and 'black-soil'); life form of vegetation upper stratum (grass, shrub, woodland and forest) and riparian or non-riparian vegetation; deciduous vine thickets and mangroves were regarded as additional categories. The range of habitat types from which each mammal species was recorded (excluding bats, which were sought in few situations—caves, watercourses and mangroves) are listed in **Appendix III**, along with number of individuals captured or seen in each. In these terms *Zyzomys argurus* occupied the widest range of habitat, followed by *Pseudomys nanus, Dasyurus hallucatus, Rattus tunneyi, Zyzomys woodwardi, Isoodon macrourus*, and *Canis familiaris.* However, the number of habitats recorded as utilised by a species is obviously related to the number of individuals sighted or captured (i.e. the more abundant or obvious a species is, the more likely it is to have been recorded in each habitat in which it occurs). Thus the range of habitats recorded for some less common species might have been increased with the capture of more individuals. Further collecting may show, however, that mammals in the bottom half of **Appendix III** utilise a narrower range of the selected habitat types than others higher up the list (this is certainly probable with *Melomys cervinipes*); this may explain why many of this lower group are poorly represented in collections.

No species was confined to the deciduous vine thickets (which is not surprising considering their relatively small total area) although *Zyzomys woodwardi* and *Melomys cervinipes* were most frequently trapped in them. These thickets, with their relatively closed canopy and sheltered situations were utilised by a wide range of mammals including in addition to the 2 above species, *Isoodon macrourus, Dasyurus hallucatus, Rattus tunneyi, Mesembriomys macrurus* and *Zyzomys argurus.* Further, McKenzie *et al.* (1975) report *Isoodon auratus, Dasyurus hallucatus, Zyzomys woodwardi* and *Mesembriomys macrurus* from similar vine thickets on volcanics in Prince Regent River Reserve.

The extent and relative importance of the lateritic plateau is the most obvious difference between the Mitchell Plateau area and other parts of the north-west Kimberley. Ten species (*Macropus antilopinus, Trichosurus arnhemensis, Isoodon macrourus, Planigale maculata, Rattus tunneyi, Mesembriomys macrurus, Zyzomys argurus, Z. woodwardi, Pseudomys nanus* and *Melomys cervinipes*) were recorded on the lateritic plateau, although *Trichosurus arnhemensis* and *Planigale maculata* were only recorded from this habitat during the Mitchell Plateau surveys. Both were single records; there is no reason to suppose that any of these 10 species are confined to the lateritic surface at Mitchell Plateau because all have previously been recorded from other sorts of habitat in the Kimberley.

It was of interest to note that in October-November 1976 which is approaching the driest and hottest time of the year, *Rattus tunneyi, Zyzomys woodwardi* and *Melomys cervinipes* were clearly more abundant in riparian vegetation and deciduous vine thickets. On the other hand *Isoodon macrourus, Dasyurus hallucatus, Mesembriomys macrurus, Zyzomys argurus* and *Pseudomys nanus* did not appear to be any more numerous in areas with greater vegetation cover.

Unfortunately, cyclonic conditions in the Mitchell Plateau area in March 1977 reduced the amount of collecting at that time. Nevertheless most of the species captured during this latter trip were in habitat similar to that from which they had been collected in October-November 1976. Only *Rattus tunneyi* appeared to have expanded its distribution in March such that it was also trapped on the lateritic plateau where it had not previously been collected. Mist-netting and shooting of bats was singularly unsuccessful in March 1977 and possibly indicates that bats were dispersed over the area and not concentrated along the watercourses as they may have been in October-November 1976.

The habitat data recorded during this work are consistent with and add to data collected on previous Kimberley surveys. Information on the habitats of Kimberley populations of *Petrogale burbidgei, Trichosurus arnhemensis, Petaurus breviceps, Planigale maculata, Sminthopsis virginiae, Zyzomys* spp., *Pseudomys forresti, Melomys cervinipes, Nyctophilus arnhemensis*, and *Macroglossus minimus* has been extended.

(c) Kimberley mammals — period of births

Tyndale-Biscoe (1973) states that it is generally supposed that breeding in eutherians occurs in spring (September-November) to ensure that young are born when food is abundant. He considers that a similar relationship seems to hold for marsupials except that it is the emergence from the pouch rather than birth which occurs in spring.

Most studies on breeding seasons of Australian mammals have been carried out in temperate climates where September-November appears to be the season of most rapid vegetation growth and food for mammals is most abundant. This may not be the situation in the Kimberley where the months of September-November are at the end of the dry season when potable water is scarcest, temperatures are high and plant growth is diminished. One of us (K.F.K.) has collected plants extensively in the Kimberley and considers that the majority of flowering and fruiting occurs throughout the wet season (December-March). However, some seed in woody fruits retained on trees is available in any season. These fruits may be the result of sporadic flowering throughout the year or prolonged retention of seed following wet season flowering of e.g. *Eucalyptus* (Myrtaceae), *Buchanania* (Anarcardiaceae), *Canarium* (Burseraceae), *Owenia* (Meliaceae), *Hakea* (Proteaceae) and *Terminalia* (Combretaceae). Observations on season of birth in this report (most summarised in **Appendix IV**) and those of Kitchener (1978) make it possible to examine the statement that September-November (the end of the dry season) may not be the most suitable season in the Kimberley for birth or emergence of young from pouch. The situation appears to be as follows:

Macropodidae: *Macropus robustus, M. agilis, Onychogalea unguifera* and *Petrogale brachyotis* have young emerging from the pouch throughout the year. *M. antilopinus* have small young at heel in October; a large furred pouch young was seen in January which would probably emerge before September. *Petrogale burbidgei* has small unfurred young in June-August and late October; these October young would probably not emerge from the pouch before February.

Dasyuridae: *Dasyurus hallucatus* and *Planigale ingrami subtilissima* young probably leave the pouch from September through to February. Young of the October *Sminthopsis virginiae* probably leave the pouch about the end of the dry season. Lactating *Sminthopsis macroura* (Gould, 1845) have been recorded in October; no females are available from other seasons.

Muridae: *Rattus tunneyi* and *Pseudomys nanus* probably give birth to young throughout the year except September-November. *Conilurus penicillatus* appears not to give birth towards the end of the dry season. *Zyzomys argurus, Z. woodwardi* and *Pseudomys delicatulus* probably give birth throughout the year, although at Mitchell Plateau *Z. argurus* and *Z. woodwardi* appear to have reduced breeding at the end of the dry season. *Mesembriomys macrurus* probably gives birth to young throughout the dry season (no female specimens from the wet season) and *Melomys cervinipes* gives birth to young at the end of the dry season (no females examined from other seasons).

Chiroptera: births occur from the end of the dry season through to the middle of the wet season in *Nyctophilus arnhemensis, N. walkeri, Miniopterus schreibersii, Eptesicus pumilus, E. douglasi, Chalinolobus gouldii, C. nigrogriseus* (**Vespertilionidae**), *Hipposideros ater, H. stenotis* (**Hipposideridae**), *Taphozous georgianus* (**Emballonuridae**) and *Macroglossus minimus* (**Macroglossidae**). Apart from *T. georgianus* which continues to give birth to young through to the end of the wet season, most of the others probably cease births towards the end of the wet season. From the size of foetuses in September-November most of the above bats have their peak period of births in the wet season. *Eptesicus pumilus* (and possibly *Macroglossus minimus*) give birth to young in all seasons. *Pteropus scapulatus* gives birth to young towards the end of the wet season.

Because period of births in species of mammals may vary with climate within the Kimberley, caution must be used in interpreting statements which combine observations on species collected from different regions within the Kimberley and from different years. Despite these reservations, our limited observations indicate that September to November is not the most important period for births of eutherians and emergence from the pouch in marsupials in tropical Western Australia. This is because in the Kimberley, no species of mammal for which there are specimens available from all seasons, is known to confine birth of its young, or emergence of its young from the pouch, to this period.

At Mitchell Plateau the following 7 species of rodents were collected in large enough numbers to allow an indication of their breeding activity at the end of the dry season: *Rattus tunneyi, Mesembriomys macrurus, Conilurus penicillatus, Zyzomys argurus, Z. woodwardi* and *Pseudomys nanus*. With the exception of *M. macrurus*, these rodents were either not breeding or had considerably reduced breeding activity compared to indications gained from study of other collections in the Kimberley of their breeding at other periods. Dwyer (1975) noted similarly that in the Eastern Highlands District of Papua New Guinea, the 10 species of murid rodents he studied were active breeders during the wet season (September to May or June) but had reduced or no breeding through the dry season (May or June to August).

ACKNOWLEDGEMENTS

We thank most sincerely AMAX Exploration for the assistance extended to us at Mitchell Plateau on each of the expeditions and to ALUMAX Inc. for so generously defraying the expenses of our March 1977 expedition to the Mitchell Plateau.

We wish to acknowledge the untiring efforts and companionship at Mitchell Plateau of the following expedition participants: R.J. Hnatiuk and B.C. Haberley (June 1976); A.M. Douglas, M. Ellis, D.T. Kitchener and W.S. & J. Street (October-November 1976); J. Henry (March 1977) and K.C. Pirkopf (May 1978).

A.M. Douglas identified the stomach contents reported in this paper.

K. Koopman, American Museum of Natural History, kindly loaned us the specimens collected from the Mitchell Plateau area in 1973 by W.H. Butler. Mr Butler was also most helpful in making available his records of the area.

Ms Robin Gregg assisted with body measurements of alcohol-preserved specimens.

Mrs Maureen Wallis deftly typed her way through this manuscript and several drafts and we thank her for her skill.

APPENDIX I

Vegetation classification of Muir (1977)

LIFE FORM/HEIGHT CLASS		CANOPY COVER			
		DENSE d 70-100%	MID-DENSE c 30-70%	SPARSE i 10-30%	VERY SPARSE r 2-10%
T	Trees >30m	Dense Tall Forest	Tall Forest	Tall Woodland	Open Tall Woodland
M	Trees 15-30m	Dense Forest	Forest	Woodland	Open Woodland
LA	Trees 5-15m	Dense Low Forest A	Low Forest A	Low Woodland A	Open Low Woodland A
LB	Trees <5m	Dense Low Forest B	Low Forest B	Low Woodland B	Open Low Woodland B
KT	Mallee tree form	Dense Tree Mallee	Tree Mallee	Open Tree Mallee	Very Open Tree Mallee
KS	Mallee shrub form	Dense Shrub Mallee	Shrub Mallee	Open Shrub Mallee	Very Open Shrub Mallee
S	Shrubs >2m	Dense Thicket	Thicket	Scrub	Open Scrub
SA	Shrubs 1.5-2.0m	Dense Heath A	Heath A	Low Scrub A	Open Low Scrub A
SB	Shrubs 1.0-1.5m	Dense Heath B	Heath B	Low Scrub B	Open Low Scrub B
SC	Shrubs 0.5-1.0m	Dense Low Heath C	Low Heath C	Dwarf Scrub C	Open Dwarf Scrub C
SD	Shrubs 0.0-0.5m	Dense Low Heath D	Low Heath D	Dwarf Scrub D	Open Dwarf Scrub D
P	Mat plants	Dense Mat Plants	Mat Plants	Open Mat Plants	Very Open Mat Plants
H	Hummock Grass	Dense Hummock Grass	Mid-Dense Hummock Grass	Hummock Grass	Open Hummock Grass
GT	Bunch grass >0.5m	Dense Tall Grass	Tall Grass	Open Tall Grass	Very Open Tall Grass
GL	Bunch grass <0.5m	Dense Low Grass	Low Grass	Open Low Grass	Very Open Low Grass
J	Herbaceous spp.	Dense Herbs	Herbs	Open Herbs	Very Open Herbs
VT	Sedges >0.5m	Dense Tall Sedges	Tall Sedges	Open Tall Sedges	Very Open Tall Sedges
VL	Sedges <0.5m	Dense Low Sedges	Low Sedges	Open Low Sedges	Very Open Low Sedges
X	Ferns	Dense Ferns	Ferns	Open Ferns	Very Open Ferns
	Mosses, liverwort	Dense Mosses	Mosses	Open Mosses	Very Open Mosses

APPENDIX II

Descriptions of trapline sites nos 1-60. These include a key description of the vegetation structure and substrate, followed by the prominent plant species in each vegetation stratum.

Trapline 1 (14°53′40″S, 125°45′20″E) ± 10″

Open Low Woodland A over Open Scrub over Hummock grass, on rugged sandstone, 150 m from Camp Creek (potable water). Stratum 1: *Planchonia australis, Xanthostemon paradoxus, Buchanania obovata, Eucalyptus brachyandra.* Stratum 2: *Calytrix* sp., *Bossiaea bossiaeoides, Acacia tumida.* Stratum 3: *Triodia* sp.

Trapline 2 (14°53′40″S, 125°45′30″E) ± 10″

Open Low Woodland A over Thicket over Hummock Grass, on rugged sandstone, 200 m from Camp Creek. Stratum 1: *Planchonia australis, Xanthostemon paradoxus, Buchanania obovata, Terminalia* sp., *Eucalyptus brachyandra.* Stratum 2: *Acacia kelleri, A. tumida, Distichostemon hispidulus.* Stratum 3: *Triodia* sp. and *Capparis lucida.*

Trapline 3 (14°53′40″S, 125°45′40″E), **4** (14°53′30″S, 125°45′50″E), **6** (14°52′45″S, 125°48′30″E), **7** (14°52′00″S, 125°49′40″E), **8** (14°52′00″S, 125°49′45″E), **11** (14°52′30″S, 125°50′00″E) and **12** (14°52′30″S, 125°50′00″E) - all ± 10″

Open Low Woodland A over Thicket over Dense Tall Grass, on edge of Camp Creek (potable water) on volcanics. Stratum 1: *Eucalyptus* sp., *Timonius timon, Planchonia australis.* Stratum 2: *Pandanus* sp., *Antidesma ghaesembilla.* Stratum 3: *Eriachne pauciflora, E. festucacea, Sorghum plumosum.*

Trapline 5 (14°53′15″S, 125°46′05″E) ± 10″

Deciduous vine thicket on volcanics. (See Beard 1976 and Hnatiuk & Kenneally (this publication) for description).

Trapline 9 (14°52′05″S, 125°49′50″E) ± 5″

Low Woodland A over Open Dwarf Scrub D over Dense Low Grass, on volcanics. Stratum 1: *Eucalyptus* spp. Stratum 2: *Eucalyptus* spp. suckering to 0.5 m. Stratum 3: *Cymbopogon bombycinus, Themeda australis.*

Trapline 10 (14°52′15″S, 125°49′50″E) ± 5″

Low Woodland A over Dense Tall Grass, on volcanics. Stratum 1: *Eucalyptus miniata, Planchonia australis, Livistona eastonii.* Stratum 2: grasses.

Trapline 13 (14°53′00″S, 125°50′10″E) ± 10″

Low Woodland B over Dense Low Grass, on volcanics. Stratum 1: *Eucalyptus* sp., *Terminalia canescens* (burnt and suckering). Stratum 2: *Heteropogon contortus, Cymbopogon bombycinus, Themeda australis.*

Trapline 14 (14°53′15″S, 125°50′00″E) ± 15″

Low Forest A over Low Woodland B over Low Grass, on volcanics. Stratum 1: *Eucalyptus miniata, E. confertiflora.* Stratum 2: *Livistona eastonii.* Stratum 3: *Heteropogon contortus, Cymbopogon bombycinus, Themeda australis.*

Trapline 15 (14°53′20″S, 125°49′40″E) ± 15″, and **17** (14°53′20″S, 125°49′50″E) ± 15″

Deciduous vine thicket, on volcanics (see Beard 1976 and Hnatiuk & Kenneally (this publication) for description).

Trapline 16 (14°53′20″S, 125°49′40″E) ± 15″

Dense Tall Grass (*Sorghum* sp.) on volcanics.

Trapline 18, 19 & 20 (14°47′20″S, 125°49′00″E) ± 20″, **57** (14°48′05″S, 125°49′20″E) ± 10″, **58** (14°43′15″S, 125°47′10″E) ± 10″, **59** (14°45′20″S, 125°47′00″E) and **60** (14°51′00″S, 125°49′50″E) ± 10″

Low Woodland A over Low Woodland B on laterite. Stratum 1: *Eucalyptus miniata, E. tetrodonta.* Stratum 2: *Livistona eastonii.*

Trapline 21 (14°46′30″S, 125°49′00″E) ± 5″

Dense Low Forest A over Dense Tall Grass, on black loam, in swamp. Stratum 1: *Melaleuca leucadendron.* Stratum 2: *Eriachne pauciflora, Sorghum aff. australiense, Pseudopogonatherum contortum, Panicum trachyrhachis.*

Trapline 22 (14°40′20″S, 125°44′00″E) ± 10″

Low Woodland A over Open Tall Grass, on sandstone on edge of creek (with potable water). Stratum 1: *Melaleuca leucadendron, Eucalyptus* sp., *Pandanus* sp. Stratum 2: *Eriachne pauciflora, Coelorachis rottboellioides.*

Trapline 23 (14°30′30″S, 125°43′50″E) ± 10″

Open Low Woodland B over Open Low Scrub A over Hummock Grass, on flat sandstone. Stratum 1: *Eucalyptus brachyandra, Terminalia* sp. Stratum 2: *Acacia orthocarpa, Atalaya simplicifolia.* Stratum 3: *Triodia* sp. with patches of *Sorghum* sp.

Trapline 24 (14°40′30″S, 125°43′40″E) ± 10″

Open Low Woodland B over Open Low Scrub A over Hummock Grass, on rugged sandstone along creek slope. Stratum 1: *Owenia vernicosa, Eucalyptus brachyandra, Terminalia* sp. Stratum 2: *Acacia orthocarpa, Atalaya simplicifolia.* Stratum 3: *Triodia* sp.

Trapline 25 (14°40′40″S, 125°43′30″E) ± 15″

Woodland over Dwarf Scrub Cover Very Open Tall Grass, on flat sandstone. Stratum 1: *Eucalyptus* sp., *Terminalia* sp. Stratum 2: *Acacia* sp., *Bossiaea bossiaeoides.* Stratum 3: *Heteropogon contortus, Sorghum* sp., *Fimbristylis* sp.

Trapline 26 (14°40′20″S, 125°43′40″E) ± 15″

Deciduous vine thicket, on sandstone scree slope by Surveyors Pool (see also Beard, 1976). Stratum 1: *Melaleuca leucadendron, Ficus platypoda.* Stratum 2: *Gardenia* sp., *Pouteria sericea, Calytrix brachychaeta.* Stratum 3: *Triodia* sp., *Mimusops elengi.*

Trapline 27 (14°40′20″S, 125°44′10″E) ± 10″

Low Woodland A over Low Forest B over Tall Grass, on black soil 100 m from creek. Stratum 1: *Eucalyptus papuana.* Stratum 2: *Melaleuca viridiflora.* Stratum 3: *Sorghum* sp., *Eriachne pauciflora.*

Trapline 28 (14°39′40″S, 125°46′20″E) ± 15″

Low Woodland A over Low Scrub B over Dense Tall Grass on volcanics, near creek with potable water. Stratum 1: *Celtis phillippensis, Terminalia canescens, Ficus platypoda* and *Pandanus* sp. Stratum 2: *Acacia stigmatophylla, Albizia lebbek.* Stratum 3: *Heteropogon contortus, Eriachne pauciflora, Cyperus* sp.

Trapline 29 (14°39′50″S, 125°46′30″E) ± 20″

Low Forest A over Low Scrub B over Open Low Grass, on volcanics near dry watercourse. Stratum 1: *Celtis phillippensis, Terminalia canescens, Ficus platypoda.* Stratum 2: *Acacia stigmatophylla, Albizia lebbek.* Stratum 3: *Heteropogon contortus, Germainia truncatiglumis, Cyperus* sp.

Traplines 30 & 31 (14°39′30″S, 125°47′00″E) ± 40″

Deciduous vine thicket, on western edge of lateritic plateau (see also Beard 1976). Stratum 1: *Bombax ceiba* var. *leiocarpum,* up to 20 m tall, sparse. Stratum 2: *Eucalyptus papuana, Clerodendrum* sp., *Mimusops elengi, Pouteria sericea, Albizia lebbek,* 10-12 m tall. Stratum 3: *Gardenia* sp., *Wrightia pubescens, Capparis umbonata.* Stratum 4: *Asparagus racemosus, Flagellaria indica.*

Trapline 32 (14°39′30″S, 125°47′00″E) ± 40″

Low Forest A over Dwarf Scrub D, on red loam soil, on edge of lateritic plateau. Stratum 1: *Eucalyptus confertiflora.* Stratum 2: *Acacia bidwillii.*

Trapline 33 (14°39′30″S, 125°46′00″E) ± 40″

Low Woodland A over Open Low Woodland B over Open Low Grass, on black soil, amongst volcanics. Stratum 1: *Eucalyptus* spp. Stratum 2: *Antidesma ghaesembilla.* Stratum 3: *Sorghum* sp., *Eriachne pauciflora.*

Trapline 34 (14°40′10″S, 125°43′50″E) ± 20″

Low Woodland A over Scrub over Hummock Grass, on rugged sandstone. Stratum 1: *Terminalia* sp., *Planchonia australis*. Stratum 2: *Grevillea heliosperma*. Stratum 3: *Triodia* sp.

Trapline 35 (14°40′10″S, 125°44′20″E) ± 20″

Open Low Woodland A over Open Low Grass, on black soil. Stratum 1: *Eucalyptus apodophylla, Melaleuca viridiflora, Pandanus* sp. Stratum 2: *Pseudopogonatherum contortum, Eriachne pauciflora.*

Trapline 36 (14°39′30″S, 125°47′00″E) ± 40″

Low Forest A over Open Low Woodland A over Dwarf Scrub D, on edge of lateritic plateau. Stratum 1: *Eucalyptus tetrodonta, E. miniata*. Stratum 2: *Acacia pachyphloia, Livistona eastonii*. Stratum 3: regenerating *Eucalyptus tetrodonta* and *E. miniata* suckers.

Trapline 37 (14°36′30″S, 125°52′00″E) ± 15″

Deciduous vine thicket, between beach and volcanic cliff (see also Beard 1976). Stratum 1: *Melaleuca leucadendron, Ficus platypoda, Terminalia petiolaris, Albizia lebbek*. Stratum 2: *Gardenia* sp., *Pouteria sericea, Calytrix brachychaeta, Bombax* sp. Stratum 3: *Flagellaria indica, Erythrophleum chlorostachys.*

Trapline 38 (14°36′30″S, 125°52′10″E) ± 15″

Low Woodland A over Open Low Woodland B over Open Low Scrub A over Tall Grass, on gently sloping volcanic soils below cliffs. Stratum 1: *Melaleuca leucadendron, Ficus platypoda*. Stratum 2: *Bombax ceiba* var. *leiocarpum, Grevillea* sp., *Terminalia canescens, Hakea arborescens*. Stratum 3: *Jasminum* sp., *Capparis umbonata.*

Trapline 39 (14°36′40″S, 125°52′10″E) ± 15″

Open Low Woodland B over Open Low Scrub A over Open Tall Grass over Open Low Grass, on volcanics near creek with potable water. Stratum 1: *Cochlospermum fraseri, Terminalia canescens, Pandanus* sp. Stratum 2: *Tephrosia* sp., *Lysiphyllum cunninghamii*. Stratum 3: *Sorghum* sp., *Eriachne pauciflora*. Stratum 4: *Bulbostylis* sp.

Trapline 40 (14°37′00″S, 125°51′30″E) ± 20″

Semi-deciduous vine thicket on volcanics. Stratum 1: *Bombax ceiba* var. *leiocarpum, Lysiphyllum cunninghamii, Erythrophleum chlorostachys, Celtis philippensis*. Stratum 2: *Capparis umbonata*. Stratum 3: *Flagellaria indica.*

Trapline 41 (14°36′30″S, 125°51′40″E) ± 15″

Interphase between deciduous vine thicket and Low Woodland A over Low Woodland B over Low Scrub, on volcanic scree above beach. (Description of vine thicket as for trapline 37.) Stratum 1: *Eucalyptus* spp. Stratum 2: *Bauhinia* sp., *Lysiphyllum cunninghamii, Terminalia* sp. Stratum 3: *Thespesia populneoides, Flagellaria indica.*

Trapline 42 (14°36′50″S, 125°51′10″E) ± 10″

Interphase between deciduous vine thicket and mangrove. Vine thicket on basaltic scree slope separated from mangrove by a beach ca. 15 m wide. A small creek flows through mangrove of *Aegiceras corniculatum, Avicennia marina* and *Xylocarpus australasicus* (less common). (Description of vine thicket as for trapline 37.)

Trapline 43 & 45 (14°33′30″S, 125°50′30″E) ± 20″

Deciduous vine thicket, on volcanic scree. (Description as for trapline 37.)

Trapline 44 (14°33′30″S, 125°50′30″E) ± 20″

On beach adjoining deciduous vine thickets of traplines 43 & 45.

Trapline 46 (14°33′35″S, 125°50′40″E) ± 20″

Low Forest A over Dense Tall Grass, on volcanic slope. Stratum 1: *Eucalyptus miniata, E. tetrodonta, Bombax ceiba* var. *leiocarpum, Bauhinia* sp. Stratum 2: *Sorghum australiense, Cenchrus elymoides.*

Trapline 47 & 48 (14°34′00″S, 125°50′30″E) ± 10″

Open Low Woodland A over Open Tall Grass over Tall Sedges on volcanic slope on edge of watercourse with potable water. Stratum 1: *Eucalyptus miniata, E. tetrodonta*. Stratum 2: *Sorghum* sp. Stratum 3: sedges.

Trapline 49 (14°33′55″S, 125°49′40″E) ± 10″

Interphase between mangrove and Open Low Woodland A. The mangrove was predominately *Avicennia marina* and *Lumnitzera racemosa,* both to 3 m. The Open Low Woodland was very similar to that described for trapline 51.

Trapline 50 (14°34′00″S, 125°49′20″E) ± 10″

Open Low Woodland A over Open Low Woodland B over Dense Tall Sedges over Low Sedges, on volcanics at edge of watercourse with potable water. Stratum 1: *Dolichandrone* sp. Stratum 2: *Pandanus* sp. Stratum 3: sedges. Stratum 4: sedges, *Fimbristylis* sp.

Trapline 51 (14°33′40″S, 125°49′30″E) ± 10″

Open Low Woodland A over Dense Tall Grass, on volcanic scree slope. Stratum 1: *Eucalyptus* spp. Stratum 2: *Sorghum* sp.

Trapline 52 (14°30′00″S, 125°47′20″E) ± 20″

Low Scrub A over Hummock Grass on rugged sandstone at edge of Crystal Creek. Stratum 1: *Acacia orthocarpa.* Stratum 2: *Plectrachne* sp.

Trapline 53 (14°30′00″S, 125°47′20″E) ± 20″

Description as for line 52 but very recently burnt.

Trapline 54 (14°30′00″S, 125°47′10″E) ± 20″

Low Woodland A over Thicket over Very Open Low Grass, on black soil, 100 m from Crystal Creek. Stratum 1: *Eucalyptus* spp., *Bombax* sp. Stratum 2: *Acacia translucens.* Stratum 3: *Eriachne pauciflora.*

Trapline 55 (14°30′25″S, 125°46′40″E)

Forest over Low Woodland A over Scrub over Tall Grass, on edge of rugged sandstone along Crystal Creek. Stratum 1: *Melaleuca leucadendron.* Stratum 2: *Pandanus* sp., *Timonius timon, Dodonaea platyptera.* Stratum 3: *Acacia* sp. Stratum 4: *Sorghum* sp., *Chrysopogon latifolius.*

Trapline 56 (14°30′15″S, 125°46′30″E)

Low Woodland A over Low Woodland B over Hummock Grass, on rugged sandstone. Stratum 1: *Owenia vernicosa, Vitex glabrata, Eucalyptus* spp., *Xanthostemon paradoxus, Buchanania obovata.* Stratum 2: *Gardenia megasperma, Petalostigma pubescens, Grevillea refracta.* Stratum 3: *Tephrosia* sp., *Tribulus angustifolius.*

APPENDIX III

Number of habitats (see p. 158) utilised by mammals at Mitchell Plateau in October-November 1976. Number of individuals captured or seen are listed. Total number of habitat types recognised was 19.

Species	No. of habitats utilised	No. of individuals
Zyzomys argurus	12	93
Pseudomys nanus	7	36
Zyzomys woodwardi	6	32
Dasyurus hallucatus	6	16
Rattus tunneyi	6	22
Isoodon macrourus	6	10
Canis familiaris	6	7
Mesembriomys macrurus	4	7
Macropus antilopinus	3	7
Conilurus penicillatus	3	9
Melomys cervinipes	3	26
Petrogale burbidgei	2	3
Antechinus cf. *bilarni*	2	3
Macropus agilis	2	2
Sminthopsis virginiae	2	2
Petaurus breviceps	1	2
Macropus robustus	1	2
Petrogale brachyotis	1	1
Pseudomys forresti	1	1
Pseudomys sp.	1	1

APPENDIX IV

Number of young (from counts of foetuses, uterine scars, swellings, or nursing young) and season of birth for some Mitchell Plateau species; based on examination of Western Australian Museum Kimberley specimens.

(-) apparent absence of births
(x) births
(?) too few specimens available for comment.

Species	Number of young			Period of birth			
	mean	range	sample (N)	Dec.-Feb.	Mar-May	June-Aug.	Sept-Nov.
PETAURIDAE							
Petaurus breviceps	2	-	2	?	x	x	?
DASYURIDAE							
Dasyurus hallucatus	6	2-8	3	-	-	x	x
Sminthopsis virginiae	7	5-8	2	?	?	?	x
MURIDAE							
Rattus tunneyi	7	3-10	10	x	x	x	-
Mesembriomys macrurus	2	1-3	4	?	?	x	x
Conilurus penicillatus	3	2-4	6	?	x	x	-
Zyzomys argurus	2	1-3	21	x	x	x	x
Z. woodwardi	2	2-3	10	x	x	x	x
Pseudomys delicatulus	3	1-4	5	x	x	x	x
P. nanus	4	3-4	2	x	x	x	-
Melomys cervinipes	2	2-2	5	?	?	?	x
VERPERTILIONIDAE							
Nyctophilus arnhemensis	1	-	1	x	?	?	x
N. walkeri	2	-	1	?	?	-	x
Miniopterus schreibersii	1	-	5	x	?	-	x
Eptesicus pumilus	2	1-2	10	x	x	x	x
E. douglasi	1	-	1	?	-	-	x
HIPPOSIDERIDAE							
Hipposideros stenotis	1	-	2	x	?	-	x
EMBALLONURIDAE							
Taphozous georgianus	1	-	51	x	x	-	x
PTEROPODIDAE							
Macroglossus minimus	1	-	9	?	?	?	x

REFERENCES

ALLEN, A.D. (1971)—*Explanatory notes on the Montague Sound geological sheet area.* Canberra: Bureau of Mineral Resources.

ANON. (1974)—C.S.I.R.O. *Rep. Div. Wildl. Res. C.S.I.R.O. Aust.* **1972-74:** 9-10.

ARCHER, M. (1979)—Revision of the marsupial dasyurid genus *Sminthopsis* Thomas, a description of two new species from northern Australia, *S. butleri* and *S. douglasi*, and a consideration of arid-adapted characters. *Bull. Am. Mus. nat. Hist.* [In press.]

BEARD, J.S. (1976)—Monsoon forest of the Admiralty Gulf, Western Australia. *Vegetatio.* **31:** 177-192.

CALABY, J.H. & KEITH, K. (1974)—Mammals. *In*: Fauna survey of Port Essington District, Cobourg Peninsula, Northern Territory of Australia. (Eds H.J. Frith & J.H. Calaby) *C.S.I.R.O. Aust. Div. Wildl. Res.* Tech. Pap. No. 28.

CRICHTON, E.G. (1969)—Reproduction in the pseudomyine rodent *Mesembriomys gouldii* (Gray) (Muridae). *Aust. J. Zool.* **5:** 785-799.

DAVIES, S.J.J.F. (1960)—A note on two small mammals of the Darwin area. *J. R. Soc. West. Aust.* **43:** 63-6.

DWYER, P.D. (1963)—The breeding biology of *Miniopterus schreibersii blepotis* (Temminck) (Chiroptera)in north-eastern New South Wales. *Aust. J. Zool.* **11:** 219-40.

— (1970)—Foraging behaviour of the Australian large-footed *Myotis* (Chiroptera). *Mammalia* **34:** 76-80.

— (1975)—Observations on the breeding biology of some New Guinea murid rodents. *Aust. Wildl. Res.* **2:** 33-45.

FRITH, H.J. & CALABY, J.H. (1969)—*Kangaroos.* New York: Humanities Press. 209 pp.

GOODING, C.D. (1963)—Wallaby control in the Kimberleys. *J. Agric. West Aust.* (4) **4:** 611-613.

GOULD, J. (1851)—Text to plate 5, *Hapalotis penicillata.* In: *The mammals of Australia.* Vol. 3. London: J. Gould.

JOHNSON, D.H. (1964)—Mammals of the Arnhem Land Expedition. Pp. 427-515 in: *Records of the American-Australian Scientific Expedition to Arnhem Land.* **4:** Ed. R.L. Specht. Melbourne: Melbourne Univ. Press.

KITCHENER, D.J. (1973)—Reproduction in the common sheath-tailed bat, *Taphozous georgianus* (Thomas) (Microchiroptera: Emballonuridae), in Western Australia. *Aust. J. Zool.* **21:** 375-89.

— (1976)—*Eptesicus douglasi*, a new vespertilionid bat from Kimberley, Western Australia. *Rec. West. Aust. Mus.* **4:** 295-301.

— (1978)—Mammals of the Ord river area, Kimberley, Western Australia. *Rec. West. Aust. Mus.* **6:** 189-219

KITCHENER, D.J. & SANSON, G. (1978)—*Petrogale burbidgei* (Marsupiala, Macropodidae), a new rock wallaby from Kimberley, Western Australia. *Rec. West. Aust. Mus.* **6:** 269-285.

McKEAN, J.L. & HALL, L.S. (1965)—Distribution of the large-footed myotis, *Myotis adversus* in Australia. *Victorian Nat.* **82:** 164-8.

McKEAN, J.L. (1972)—Notes on some collections of bats (order Chiroptera) from Papua-New Guinea and Bougainville Island. *Tech. Pap. Div. Wildl. Res. C.S.I.R.O. Aust.* no. 26.

McKENZIE, N.L., CHAPMAN, A. & YOUNGSON, W.K. (1975)—Mammals of the Prince Regent River Reserve, north-west Kimberley, Western Australia. Wildl. Res. Bull. West. Aust. no. 3: 69-74.

McKENZIE, N.L., CHAPMAN, A., YOUNGSON, W.K. & BURBIDGE, A.A. (1977)—Mammals of the Drysdale River National Park, north Kimberley, Western Australia. *Wildl. Res. Bull. West. Aust.* no. 6: 79-85.

McKENZIE, N.L., BURBIDGE, A.A., CHAPMAN, A. & YOUNGSON, W.K. (1978)—Mammals. *Res. Bull. West. Aust.* no. 7: 22-28.

MARLOW, B.J. (1958)—A survey of the marsupials of New South Wales. *C.S.I.R.O. Wildl. Res.* **3:** 71-114.

MUIR, B.G. (1977)—Vegetation and habitat of Bendering Reserve. Biological Survey of the Western Australian wheatbelt. Pt. 2. *Rec. West. Aust. Mus.* Suppl. no. 3.

NELSON, J.E. (1965)—Movement of Australian flying foxes (Pteropodidae: Megachiroptera). *Aust. J. Zool.* **13:** 53-73.

NEWSOME, A.E., CORBETT, L.K., BEST, L.W. & GREEN, B. (1973)—The dingo. *A.M.R.C. Review* **14:** 1-11.

PARKER, S.A. (1973)—An annotated checklist of the native land mammals of the Northern Territory. *Rec. S. Aust. Mus.* **16**(11): 1-57.

RATCLIFFE, F.N. (1931)—The flying fox (*Pteropus*) in Australia. *Bull. Coun. scient. ind. Res., Melb.* no. 53.

START, A.N. (1974)—The feeding biology in relation to food sources of nectarivorous bats (Chiroptera: Macroglossinae) in Malaysia [Ph.D. Thesis: University of Aberdeen, Aberdeen].

TAYLOR, J.M. & HORNER, B.E. (1970)—Observations on reproduction in *Leggadina* (Rodentia, Muridae). *J. Mammal.* **51:** 10-17.

— (1971)—Reproduction in the Australian tree rat *Conilurus penicillatus* (Rodentia: Muridae). *C.S.I.R.O. Wildl. Res.* **16:** 1-9.

— (1973a)—Results of the Archbold Expeditions. No. 98. Systematics of native Australian *Rattus* (Rodentia: Muridae). *Bull. Am. Mus. nat. Hist.* **150:** 1-130.

— (1973b)—Reproductive characteristics of wild native Australian *Rattus* (Rodentia: Muridae). *Aust. J. Zool.* **21:** 437-75.

TYNDALE-BISCOE, H. (1973)—*Life of marsupials* Canberra: Edward Arnold.

WHITEHOUSE, S.J.O. (1977)—The diet of the dingo in Western Australia. *Aust. Wildl. Res.* **4:** 145-50.

Received 1 December 1978 Accepted 19 September 1979 Published 28 February 1981

PART 5

BIRDS OF MITCHELL PLATEAU AND ADJACENT COASTS AND LOWLANDS, KIMBERLEY, WESTERN AUSTRALIA

R.E. JOHNSTONE

and

L.A. SMITH*

INTRODUCTION

The area covered by this report occupies 4400 km² between latitudes 14°25′ and 15°08′S and longitudes 125°40′ and 126°12′E (**Fig. 1**). It is bounded by the Mitchell River in the west and south, Admiralty Gulf in the north and the upper King Edward River in the east.

The geology and climate have been described by Wilson (Part 1, this publication).

Nothing was known of the avifauna of the Mitchell Plateau area before 1970 when naturalist J.A. Smith commenced work with Amax Exploration. It was mainly through Mr Smith's efforts that the mining company allowed us to spend part of the 1972-73 wet season in the area. Data presented here were obtained by us from 14 January-24 February 1973, 18-24 June 1975 and 15 October-7 November 1976. We are grateful for unpublished data from J.A. Smith (1970-73 and 6 September 1978-28 April 1979), W.H. Butler (21 July-10 August 1973, 22-25 October 1973, 25 September-8 October, 26 October-7 November 1978 and 21 February-4 March 1979) and H.B. Gill (1-3 July 1973). The initials of these observers follow where theirs are the only data pertaining to the species.

The vegetation of the Mitchell Plateau area has been briefly studied by Beard (1976) and in detail by Hnatiuk and Kenneally (Part 2, this publication). For the purpose of this paper we recognise 7 major habitat types: the lateritic plateau (excluding Airfield Swamp); Airfield Swamp itself; semi-deciduous vine forests, thickets and scrubs; alluvial flats; woodlands on basalts; King Leopold Sandstones; and mangals.

1. Lateritic Plateau (Figs 1, 2, 3)

The plateau is covered with woodland of *Eucalyptus tetrodonta* and *E. miniata* with scattered *E. nesophila* and *E. latifolia.* The palm *Livistona eastonii*, 3-10 m high, forms a dense lower layer. The shrub layer is almost non-existent, probably the result of annual burning; the woody plants here include the small trees and shrubs

---

*Department of Ornithology and Herpetology, Western Australian Museum, Francis Street, Perth, Western Australia. 6000.

Erythrophleum chlorostachys, Terminalia canescens, Cochlospermum fraseri, Grevillea agrifolia, G. parallela and *G. refracta, Petalostigma pubescens* and *Persoonia falcata.* The cycad *Cycas lane-poolei* is found near the southern end of the plateau, and extensive areas of *Cycas basaltica* grow on the northern, eastern and southern slopes. Several small groves of *Banksia dentata* occur near the airstrip and Airfield Swamp. Grasses include *Sorghum plumosum, Themeda australis, Chrysopogon latifolius* and *Plectrachne pungens.*

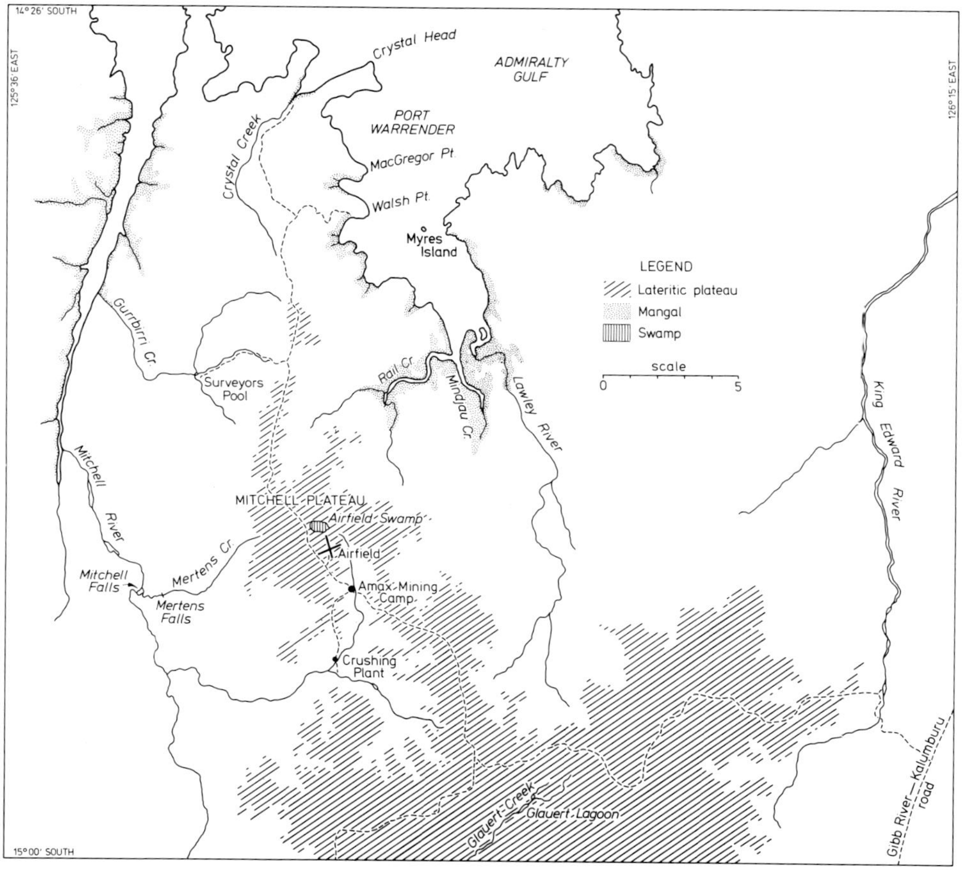

Fig. 1: Map of Mitchell Plateau region, Western Australia.

2. Airfield Swamp (Fig. 4)

This swamp is 5 km north of Mining Camp and has an area of about 80 ha **(Fig. 1)**. It is surrounded by a dry outer zone of *Sorghum plumosum* then a tall dense wet belt of *Pseudopogonatherum contortum* and *Eriachne pauciflora.* The trees comprise a pure stand of *Melaleuca leucadendron* 8-10 m tall. Sparse *Sorghum* and aquatics (including *Nymphoides* spp.) grow in the water on grey clay.

The swamp fills during the wet season and retains water through to about July. We observed it full in January-February 1973 with 1.5 m of water in the centre and flooding to the edge of the airstrip; in June 1975 and 1976 it had 1 m of water in the centre; in June 1974 Beard found 60 cm of water; in September-October 1976 it was dry.

3. Semi-deciduous vine forests, thickets and scrubs (Fig. 5)

Scattered around the edges of the lateritic plateau and on coastal basalt scree slopes are areas of open to closed semi-deciduous vine forests, some up to 40 ha but generally less than 25 ha. Vine thickets and scrubs also occur in riverine situations. The main forest areas are on the rubble-strewn talus slopes of the plateau. These forests are mainly stunted formations in which the canopy closes at 3-9 m with mostly deciduous emergents up to 15 m. The emergents include *Bombax ceiba* var. *leiocarpum, Albizia lebbek, Celtis philippensis, Randia cochinchinensis, Ficus* spp., *Alstonia, Eugenia* spp., *Vitex glabrata* and *Ganophyllum falcatum*. The lower level consists mostly of slender trees and shrubs including *Terminalia* spp., *Exocarpos, Diospyros nitens* and *Atalaya*. The lowest level is the spiny shrub layer, so called because of the dominance of *Zizyphus* spp., *Micromelum, Mellochia umbellata, Citriobatus spinescens* and *Capparis*; the vines *Abrus precatorius* and *Flagellaria indica* scrambling over the trees and rocks make the layer almost impenetrable. The ground is thickly carpeted with fallen leaves.

The coastal thickets on the scree slopes are generally more open, with large trees of *Ficus* spp., *Bombax, Albizia, Zizyphus, Terminalia, Myristica, Canarium, Mimusops* etc. The riverine scrubs and thickets consist mostly of *Ficus, Terminalia, Tristania, Barringtonia* and *Pandanus*.

4. Alluvial Flats (Fig. 8)

These areas of black soil occur between the sandstones and basalts. They support a low woodland of *Eucalyptus apodophylla, E. latifolia, E. bigalerita, E. papuana, E. polycarpa* and *Terminalia fitzgeraldii*, with an understory of smaller trees and shrubs. The ground cover is mostly a tall dense layer of grasses *Sorghum* and *Eriachne*. Waters in this area include the lagoons 20 km south-south-east of Mining Camp. They are situated in open eucalypt woodland with scattered stands of *Banksia dentata* and *Grevillea pteridifolia* and lined with *Pandanus*.

5. Woodlands on Basalts (Fig. 6, 7)

These are located on the slopes of the plateau. The vegetation is a savannah woodland of *Eucalyptus tectifica, E. foelscheana, E. confertiflora*, scattered *E. miniata*, and an understorey of *Acacia, Grevillea* and *Hakea*. The ground cover is mainly tall grasses.

6. King Leopold Sandstones (Fig. 9)

The heavily dissected sandstones about Mitchell River and Surveyors and Crystal Creeks are clothed with low woodland, tall shrubland or tall open shrubland. Trees and shrubs include *Eucalyptus miniata, E. brachyandra, E. herbertiana*, bloodwoods (*Eucalyptus* spp.), *Owenia vernicosa, Acacia* spp. (including *A. holosericea* and *A. kelleri*), *Ficus* spp. (including *F. virens* var. *sublanceolata* and *F. platypoda*), *Brachychiton* spp., *Verticordia cunninghamii* and *Xanthostemon paradoxus*.

Ground cover is mainly spinifex (*Triodia pungens*). Along creeks and rivers are areas of alluvial sand with a eucalypt woodland, mainly of *E. miniata, E. tetrodonta, Banksia dentata,* and *Pandanus*, with scattered *Grevillea pteridifolia* and *Verticordia cunninghamii* and a dense ground cover of grasses.

The Mitchell, Lawley and King Edward Rivers and Surveyors and Crystal Creeks are fringed with forest of varying composition. Tall *Melaleuca leucadendron* and *M. argentea* are common with dense stands of *Pandanus*, and in the more rocky places a tall shrubland of *Eugenia, Terminalia, Ficus, Barringtonia* and *Calytrix.* The sandstones east of the plateau are flatter and less rugged, but their vegetation is fairly similar to that of the dissected western areas.

7. Mangals (Fig. 10)

Port Warrender includes many broad shallow bays which contain extensive stands of mangroves. In Port Warrender we mainly worked south of Walsh Point where the mangal has distinct zonation. Here it consisted of a pioneer or seaward zone of *Avicennia marina* and *Sonneratia alba*, a middle zone of low closed forest of *Rhizophora stylosa* and *Camptostemon schultzii*, and a landward zone of *Bruguiera* spp. and *Ceriops tagal.*

The other mangal investigated was along Rail Creek (**Fig. 1**). The mangal here is not as well zoned but is richer floristically, because of the varied physiography: extensive shelving mudbanks, numerous small creeks running into the mangal and sandy, muddy and rocky substrates. The dominant creek vegetation is a mixed low closed forest of *Avicennia marina, Camptostemon schultzii, Xylocarpus granatum, Rhizophora stylosa, Sonneratia alba, Bruguiera exaristata, B. parviflora, Aegiceras corniculatum* and with often a ground cover of *Aegialitis* especially on the seaward side. There are dense stands of *Ceriops tagal* on the landward fringe and tall pure stands of *Bruguiera parviflora* on the upper reaches of Rail Creek. Whipstick thickets of *Aegiceras* often occur in the seaward zone among *Rhizophora* etc.

Brief visits were also made to the mangals at the mouth of the Lawley River, Mindjau and Crystal Creeks. The vegetation at the mouth of Crystal Creek is mainly a mixed low closed forest of *Sonneratia, Avicennia, Rhizophora, Camptostemon, Bruguiera* and *Ceriops.*

In the annotated list we summarize for each species its relative abundance, its habitat preferences, and where available, data on breeding and notes on taxonomy. The 205 specimens collected are lodged in the Western Australian Museum under registered numbers A12317-A12322, A12325-A12444, A14005-A14009, A14015-A14018, A14020-A14021, A14023, A14355-A14356, A14377-A14379, A14381-A14382, A14560-A14563, A14654-A14688, A14690-A14708, A15154, A15290. The 81 specimens collected by W.H. Butler in September, October and November 1978 and February-March 1979 are lodged in the Western Australian Museum under registered numbers A15734-A15785 and A15929-A15955.

The classification and nomenclature follow those of Storr (1980).

Fig. 2: Lateritic plateau showing *Eucalyptus tetrodonta* and *Livistona eastonii.*

Fig. 3: Lateritic plateau after burn.

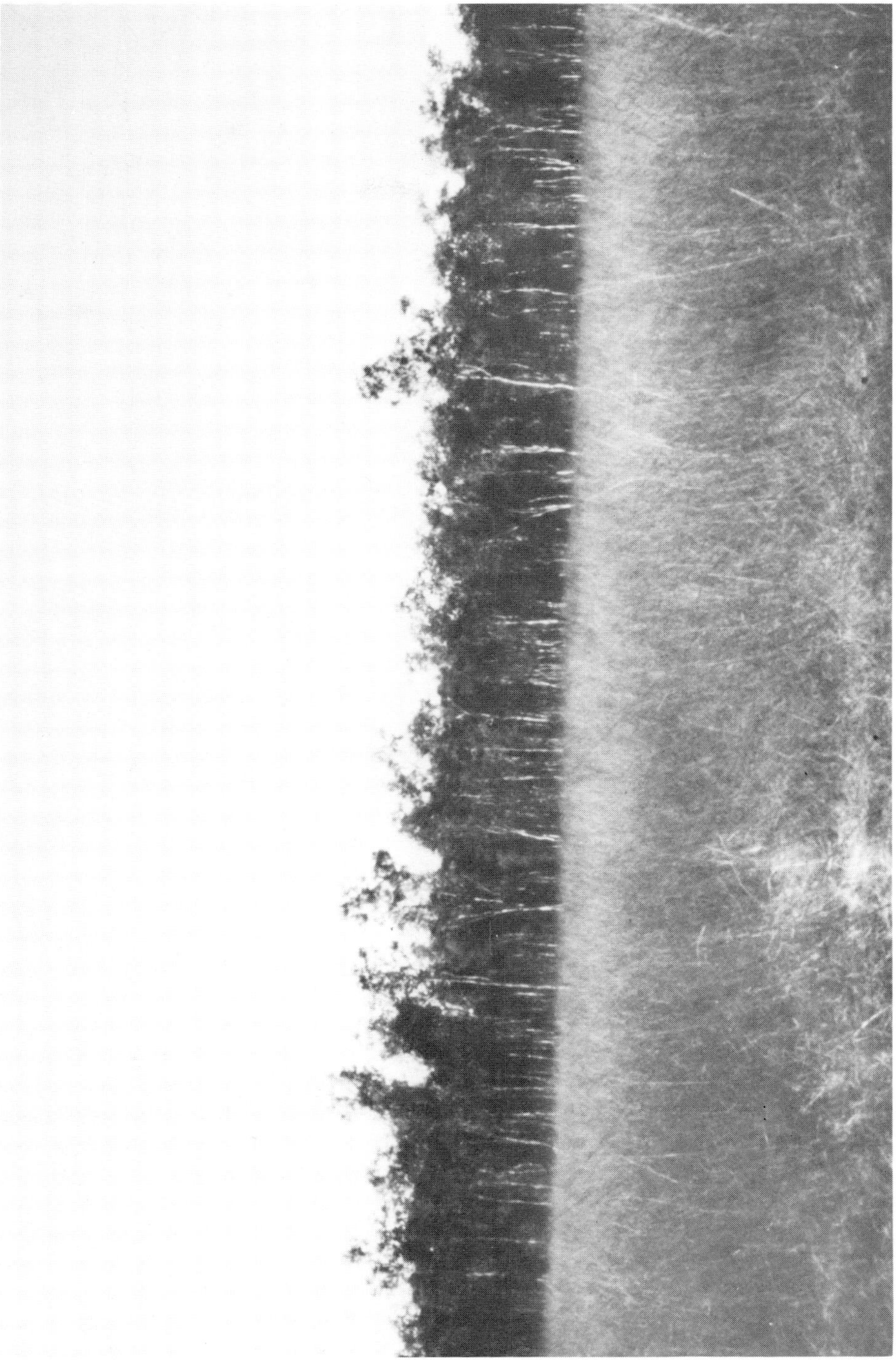

Fig. 4: Airfield Swamp.

Fig. 5: Semi-deciduous vine forest at edge of plateau.

Fig. 6: Woodland on basalt with a small patch of vine thicket in background.

Fig. 7: Tall grassland on basalt near Port Warrender.

Fig. 8: Glauert Lagoon.

Fig. 9: King Leopold Sandstones near Mitchell Falls.

Fig. 10: Rail Creek showing mangal, open woodland and some deciduous emergents from small vine thicket in foreground.

ANNOTATED LIST OF SPECIES

CASUARIIDAE

Dromaius novaehollandiae (Latham) **Emu**

Moderately common on plateau and alluvial flats, in ones, twos or small parties. Mainly in palm savannah with *Livistona eastonii, Eucalyptus tetrodonta* and *E. miniata.* Birds noted as being very dark. Adult with 3 chicks 0.6 m high on 18 October 1970; adult with 2 chicks about 1 month old on 18 October 1971; adult with 8 chicks a few days old on 6 August 1972; and adult with 5 chicks 1 day old on 4 September 1972. On 6 September 1978 pet Emu at Mining Camp followed by 10 chicks, the oldest approximately 8 weeks old, 5 about 4 weeks old and 4 about 1 to 2 weeks old. Scats contained seeds of *Petalostigma pubescens* and *Livistona eastonii.*

PODICIPEDIDAE

Podiceps novaehollandiae novaehollandiae Stephens **Black-throated Grebe**

Scarce. One on the plateau on 9 May 1971. One on Glauert Lagoon (20 km south-east of Mining Camp) on 30 October 1976. One on a large pool 5 km east of Mining Camp on 11 February 1979. One on pool 8 km east of Mining Camp on 14 January 1979.

Two nests, one with 8 eggs and the other still being built, at Airfield Swamp on 28 March 1979. Four of the 8 eggs had hatched by 7 April, the other 4 by 16 April. The second nest was built but never used, the female apparently laying in the first nest.

PELECANIDAE

Pelecanus conspicillatus Temminck **Australian Pelican**

Scarce. Two at Port Warrender on 2 July 1973 (H.B.G.), 1 on 23 September 1978, 13 on 4 March and 8 on 21 March 1979 (J.A.S.).

SULIDAE

Sula leucogaster plotus (Forster) **Brown Booby**

One at Port Warrender on 23 September 1978 (J.A.S.).

PHALACROCORACIDAE

Phalacrocorax sulcirostris (Brandt) **Little Black Cormorant**

Scarce. Five on rocks at Surveyors Pool on 6 November 1976.

Phalacrocorax varius (Gmelin) **Pied Cormorant**

Scarce. One at Port Warrender on 4 November 1976 and 1 on 23 September 1978. J.A.S. recorded 60-65 nests with young from a week old to almost fully fledged on a bare island in Admiralty Gulf on 17 July 1970.

Phalacrocorax melanoleucos (Vieillot) **Little Pied Cormorant**

Uncommon. Ones and twos on major waters (Surveyors Pool, Rail Creek, Airfield Swamp, and Glauert Lagoon).

Anhinga melanogaster novaehollandiae (Gould) **Darter**

Uncommon (8 sightings). Mostly single birds (once 2). Watercourses with cadjeput lined-pools (Camp Creek, Mitchell River), and lagoons 20 km south-east of Mining Camp, Lawley River delta and mangal south of Walsh Point.

FREGATIDAE

Fregata ariel (Gray) **Lesser Frigate-bird**

Scarce. One at Port Warrender on 27 October 1976 and one on 23 September 1978.

ARDEIDAE

Ardea sumatrana Raffles **Great-billed Heron**

Scarce. One in estuary of Lawley River on 29 August 1971; 1 on sandspit at river mouth; 1 in dense mixed mangroves at Rail Creek on 17 October 1976; 1 near Walsh Point on 14 April 1979.

Ardea pacifica Latham **Pacific Heron**

Scarce. Two at Airfield Swamp on 22 June 1975; 1 at pool on King Edward River on 24 June 1975; 1 on airstrip on 12 October 1978; 1 at pool on airstrip on 4 December 1978; 1 at King Edward River on 7 December 1978.

Ardea novaehollandiae Latham **White-faced Heron**

Uncommon. Usually single, occasionally 2 together. Mainly at large pools on major watercourses. Seen at Surveyors Pool, freshwater pool on western arm of Lawley River delta, Walsh Point and creek 5 km east of Mining Camp.

Ardea picata Gould **Pied Heron**

Scarce. Recorded between April and October 1970 (J.A.S.).

Egretta alba (Linnaeus) **Great Egret**

Scarce. Mainly single birds (once 2). Freshwater pools on plateau, King Edward River and Rail Creek, and lagoons 20 km south-east of Mining Camp.

Egretta garzetta nigripes (Temminck) **Little Egret**

Scarce. Recorded between April and October 1970, and 1 at pool on King Edward River on 7 December 1978 (J.A.S.).

Egretta sacra (Gmelin) **Eastern Reef Heron**

Common at Port Warrender, including 15 (14 dark phase) on Myres Island on 4 November 1976. Rocky and muddy shores of Walsh Point and Myres Island. Also at Lawley River delta. Overall, dark-phase birds outnumber white-phase birds about 5:1.

Egretta ibis coromanda (Boddaert) **Cattle Egret**

Scarce. Three in breeding plumage at large pool on edge of Mitchell Plateau airstrip on 27 February 1972 (J.A.S.).

Butorides striatus (Linnaeus) **Mangrove Heron**

Scarce. In mangal. One at mouth of Crystal Creek in August 1973; 2 reddish-brown birds in open *Rhizophora* and *Camptostemon* near Walsh Point on 21 June

1975; a grey bird at Walsh Point on 17 October 1976; and 2 near Walsh Point on 14 April 1979. This species often feeds well away from mangroves along rocky and muddy coasts and up creeks and rivers.

Mayr (1943) recognised 4 races in Western Australia and the Northern Territory: (1) *rogersi* (Pt. Cloates-Ashburton River), (2) possibly a subspecies from Port Hedland area, (3) *cinereus* (western Kimberley, W.A.) and (4) *stagnatilus* (N.T.). There is still some doubt on the validity of *rogersi* and *cinereus*; the Mangrove Heron appears to be patchily continuous along the Western Australian coast into the Northern Territory, and both reddish-brown and grey coloured birds occur in the Pilbara and Kimberley.

Nycticorax caledonicus hilli Mathews **Rufous Night-heron**

Widespread but scarce. Usually single, occasionally twos and threes. Mainly in tall cadjeputs and pandanus along watercourses (Surveyors Creek, Lawley River and King Edward River); also at Walsh Point.

Ixobrychus flavicollis (Latham) **Black Bittern**

Scarce. Mainly single birds. Two juveniles fully feathered, except for down on head, at Camp Creek on 29 October 1971. Nest with 2 eggs in *Melaleuca leucadendron* at Camp Creek on 27 November 1978; it contained downy young on 18 December. Nest with 3 eggs in same tree on 8 February 1979; on 19 February it contained 2 downy hatchlings and the other egg with chick just pecking its way out.

CICONIIDAE

Ephippiorhynchus asiaticus (Latham) **Black-necked Stork**

Uncommon. Usually single birds, occasionally 2 together. Mudflats and rocky shores at Port Warrender, and large pools on major watercourses; also a deep lily lagoon.

THRESKIORNITHIDAE

Threskiornis aethiopicus moluccus (Cuvier) **Sacred Ibis**

Scarce. Mainly single birds. At Airfield Swamp, along creeks on plateau, in mangal and on coastal mudflats. Often feeding along seaward edge of mangal. Also bathing in freshwater pools on Rail Creek.

Threskiornis spinicollis (Jameson) **Straw-necked Ibis**

Scarce. Twelve at Camp Creek on 27 May 1972 and 3 over Mining Camp on 28 March 1979 (J.A.S.).

Plegadis falcinellus (Linnaeus) **Glossy Ibis**

Scarce. Recorded between April and October 1970; 1 on Camp Creek on 19 November 1972; one feeding in a large still pool at airstrip on 10 and 11 February 1973; and 1 near Mitchell Falls on 5 November 1978.

Platalea regia Gould **Royal Spoonbill**

Scarce. Single birds on pools at Camp Creek and airstrip in November 1972 and January 1973. Four birds, probably of this species, on a mudflat at mouth of Lawley River on 17 October 1976.

ANATIDAE

Anseranas semipalmata (Latham) **Pied Goose**

Scarce visitor. Three on plateau on 4 October 1970; pair at Mining Camp for half an hour on 28 October 1971; 4 over airfield on 27 April 1972 and 1 at lagoons 20 km south-east of Mining Camp on 30 October 1976.

Anas superciliosa Gmelin **Black Duck**

Scarce. Three on plateau on 19 November 1972 (J.A.S.).

Anas gibberifrons gracilis Buller **Grey Teal**

Scarce. Three at pool on Camp Creek on 19 November 1972 (J.A.S.).

Aythya australis (Eyton) **Hardhead**

Scarce. Three pairs at pool on creek 5 km east of Mining Camp on 11 February 1979 (J.A.S.).

Nettapus pulchellus Gould **Green Pygmy Goose**

Scarce. One on plateau on 4 October 1970, and 2 pairs at Airfield Swamp on 16 March 1979 (J.A.S.).

ACCIPITRIDAE

Elanus caeruleus notatus Gould **Black-shouldered Kite**

Scarce. One over Mining Camp on 6 February 1972, 1 circling hills near Rail Creek on 21 October 1976, and another over open woodland near Walsh Point on 3 November 1976.

Aviceda subcristata (Gould) **Crested Hawk**

Scarce. Two birds, possibly of this species, on plateau about 20 km south-east of Mining camp, in palm savannah (*Livistona eastonii, Eucalyptus tetrodonta* and *E. miniata*) in July 1973 (H.B.G.). Two in open woodland 10 km north of Surveyors Pool on 3 December 1978; 1 near airstrip on 6 March 1979; 4 near Mining Camp on 6 April 1979; and 3 near Mitchell Falls on 8 April 1979 (J.A.S.).

Lophoictinia isura (Gould) **Square-tailed Kite**

Scarce. One in *Eucalyptus miniata* woodland at King Edward River crossing on 1 July 1973, Mitchell Falls on 23 January 1975 and 24 October 1976. One over eucalypt woodland and mangal at Rail Creek on 24 October 1976.

Hamirostra melanosternon (Gould) **Black-breasted Kite**

Scarce. Mostly in ones and twos over plateau and at Crystal Head; also a pair along Crystal Creek in May 1972, and a light brown immature over Crystal Creek on 4 November 1976.

Haliastur indus girrenera (Vieillot) **Brahminy Kite**

Moderately common in and around mangal, particularly seaward edge. Usually perched in tall dead *Avicennia* and *Sonneratia*. Nest with 2 eggs in mangal south of Walsh Point on 8 August 1971.

Haliastur sphenurus (Vieillot) **Whistling Kite**

Moderately common in all habitats except mangal. Mainly on plateau, usually in ones and twos, congregating at bush fires and around Mining Camp.

Milvus migrans affinis Gould **Black Kite**

Generally uncommon. Apart from a flock of 200 over plateau on 6 September 1978 and a flock of 100 on 17 October 1978 all other records are of single birds and small groups (up to 8). Like the preceding species, most birds recorded at Mining Camp and edge of bushfires.

Accipiter fasciatus didimus (Mathews) **Brown Goshawk**

Uncommon. Several single birds along Camp Creek and at Lawley River. Female (wing 273 mm, wt 377 g) collected in thick pandanus/semi-deciduous vine thicket on 31 July 1973; and male (wing 245 mm, wt 226 g) collected in vine thicket on 5 October 1978. Also in scrub fringing watercourse at Crystal Creek and in open woodland on basalt (*Eucalyptus tectifica* and *E. foelscheana*) near Walsh Point. Breeding on plateau on 26 October 1971.

Accipiter cirrocephalus (Vieillot) **Collared Sparrowhawk**

Uncommon; mostly on plateau. Listed by J.A. Smith among "the species that are here all the wet"; he noted 1 bird at Mitchell Falls seen with a captive Tawny Frogmouth which prevented it from flying; minutes later the same bird was seen capturing one of several bats flying about entrance of cave.

Nest with 2 eggs in tree at edge of Camp Creek on 13 November 1978. One egg hatched by 18 December and both young fledged on 8 January 1979. Nest with 3 eggs near airstrip on 27 October 1978 and a nest with 2 fledglings in sandstone at Lawley River on 10 December 1978.

Accipiter novaehollandiae novaehollandiae (Gmelin) **Grey Goshawk**

Scarce. A grey-phase bird at Crystal Creek in August 1973, another that landed momentarily in a tall *Avicennia* in dense mangal at Rail Creek on 23 October 1976, and 1 at Walsh Point on 26 November 1978.

Aquila morphnoides morphnoides Gould **Little Eagle**

Two birds on the plateau on 12 November 1970 (J.A.S.).

Aquila audax (Latham) **Wedge-tailed Eagle**

Moderately common in ones and twos. Seen over most habitats (basalts, sandstones and plateau). Bird on empty nest lined with fresh leaves on plateau on 21 May 1972.

Haliaeetus leucogaster (Gmelin) **White-breasted Sea-eagle**

Uncommon. Mainly around Walsh Point and Lawley River delta. Recorded up to 25 km inland along Mitchell River. Pair at nest with 2 half grown young at mouth of Lawley River on 1 October 1978.

Circus assimilis Jardine & Selby **Spotted Harrier**

One on plateau on 15 June 1971; another on 15 June 1973; 1 near Mitchell River homestead on 31 May 1975; 1 over airstrip on 15 January 1979; 1 on plateau on 24 April 1979.

Circus aeruginosus approximans Peale **Marsh Harrier**

A pale bird in a tall dead tree on plateau, 7 February 1973.

Pandion haliaetus cristatus (Vieillot) **Osprey**

Single birds at Crystal Creek, Walsh Point and Lawley River mouth. Nest on Myres Island on 29 August 1971 with 2 fully feathered young. Pair on Myres Island on 4 November 1976 (near nest apparently not in use).

FALCONIDAE

Falco peregrinus macropus Swainson **Peregrine Falcon**

Scarce. One at King Edward River crossing on 1 July 1973; 1 at Mitchell Falls on 23 June 1975; a pair at the Falls on 31 October 1976. One 6 km south of Crystal Creek on 4 March 1979; 1 near Port Warrender on 15 April 1979; 1 at Mining Camp on 18 April 1979.

Falco longipennis longipennis Swainson **Australian Hobby**

A male (wt 211.5 g) was collected at Crystal Creek on 10 August 1973 (W.H.B.). Pair hunting over airstrip on 25 April 1979.

Falco berigora berigora Vigors & Horsfield **Brown Falcon**

The most common raptor in area, seen hunting over all habitats. Up to 10 birds congregating around Mining Camp during burning-off. Nest with 3 eggs, built 10 m up in eucalypt, 28 October 1978.

Falco cenchroides cenchroides Vigors & Horsfield **Australian Kestrel**

Scarce. Mainly around sandstone cliffs. Recorded in all months but December. Occasionally in pairs.

MEGAPODIIDAE

Megapodius reinwardt tumulus Gould **Scrub Fowl**

Scarce. Almost confined to densest stands of semi-deciduous vine forests; mostly in those fringing plateau, but once in a coastal vine scrub at Macgregor Point. Birds calling by night and day. Most calls similar to domestic fowls'; also a soft "chick-chick". Feeding on fallen fruits and scratching and raking litter for snails and insects. When disturbed flying almost vertically to perch in canopy. Several mounds found, the largest about 10 m across and 3.5 m high.

PHASIANIDAE

Coturnix ypsilophora australis (Latham) **Brown Quail**

Locally common. Coveys up to 10 (once 20). Favouring well-grassed areas on basalt, especially banks of creeks; also on plateau and at Crystal Creek.

Five specimens (2 male, 1 female and 2 in spirit). Crop of 1 specimen contained grass seeds (*Cenchrus elymoides*).

Coturnix chinensis lineata (Scopoli) **King Quail**

A female was collected at night on airstrip on 28 February 1979 (W.H.B.).

TURNICIDAE

Turnix castanota (Gould) **Chestnut-backed Button-quail**

Locally moderately common, but generally scarce. Six in burnt woodland on basalt near Walsh Point on 21 June 1975. Eight on plateau in open palm savannah on 22 June 1975. Thirty on plateau in burnt woodland on 27 October 1978. Several quail, probably of this species, flushed from rank grass near Rail Creek in October 1976.

Three specimens (2 male, 1 female).

GRUIDAE

Grus rubicundus (Perry) **Brolga**

Moderately common. Usually in pairs or small parties (up to 17). Mostly around Airfield Swamp feeding in outer grassland zones of *Sorghum plumosum* and *Pseudopogonatherum contortum*, and among *Melaleuca leucadendron*. Also along major watercourses and occasionally in open areas on plateau. Two adults and 2 juveniles at lagoons on 30 October 1976.

RALLIDAE

Gallirallus philippensis mellori (Mathews) **Banded Landrail**

One, in tall grass on basalt at edge of camp Creek on 13 January 1973, which remained around Mining Camp for about two weeks. One along Rail Creek on 18 October 1976 in thick *Typha* 2 m high.

One specimen (male).

Porzana tabuensis (Gmelin) **Spotless Crake**

An adult with a chick in tall grass fringing a creek 5 km east of Mining Camp on 5 February 1979 and 1 at same place feeding among jumbled boulders edge of water on 11 March 1979 (J.A.S.).

Eulabeornis castaneoventris Gould **Chestnut Rail**

Moderately common. Mangal south of Walsh Point and at Lawley River delta. About 4 pairs recorded in 5 ha of mangal along Rail Creek in October 1976. Usually seen feeding along seaward edge of mangal. One female collected was eating small red crabs *Sesarma* sp. which it broke open or killed by tapping against *Rhizophora* prop roots.

Three specimens (females).

Gallinula ventralis Gould **Black-tailed Native Hen**

One at Airfield Swamp on 28 March 1979 (J.A.S.).

OTIDIDAE

Otis australis Gray **Australian Bustard**

Common. Mostly in ones and twos (one party of 4). The more open woodlands on basalt and plateau. Downy young with parent on 23 October 1971. Males displaying in October and January.

HAEMATOPODIDAE

Haematopus longirostris Vieillot **Pied Oystercatcher**

Twelve on mudflat at mouth of Lawley River on 17 October 1976 and 2 on Myres Island on 4 November 1976.

Haematopus fuliginosus opthalmicus Castelnau & Ramsay **Sooty Oystercatcher**

On Myres Island, 2 birds on 4 November 1976 with nest with 2 eggs on 29 August 1971.

CHARADRIIDAE

Vanellus miles miles (Boddaert) **Masked Plover**

Five on lawns at Mining Camp on 2 December 1978 (J.A.S.).

Pluvialis squatarola (Linnaeus) **Grey Plover**

Two on mudflat at mouth of Lawley River on 17 October 1976.

Pluvialis dominica fulva (Gmelin) **Eastern Golden Plover**

One at Mining Camp on 2 October 1978; 1 in breeding plumage at airstrip on 12 October 1978; 7 at airstrip on 17 and 18 December 1978; 3 on 30 December 1978; 3 on 15 January 1979; and 2 on mudflat near Walsh Point on 4 March 1979 (J.A.S.).

Charadrius ruficapillus Temminck **Red-capped Plover**

Scarce. Recorded in 1971. About 20 at Walsh Point on 1 July 1973; 2 on mudflat at mouth of Lawley River on 17 October 1976; 9 on mudflat near Walsh Point on 14 April 1979.

Charadrius mongolus mongolus Pallas **Mongolian Sand Plover**

Flock of 42 near Walsh Point on 2 July 1973 (H.B.G.); 2 on mudflat near Walsh Point on 4 March 1979 (J.A.S.).

Charadrius leschenaultii Lesson **Large Sand Plover**

Three on sand flats near Walsh Point on 21 February 1973 and 4 there on 2 July 1973.

Charadrius melanops Vieillot **Black-fronted Plover**

Recorded between April and October 1970. Two near Walsh Point on 2 July 1973.

Charadrius veredus Gould **Oriental Plover**

Five on plateau from 22 August to 21 September 1974 (J.A.S.). A colleague of J.A. Smith said they had been there since May. Eighteen on airstrip on 22 September 1978; 34 on 1 October 1978; 4 on 12 October 1978; 30 on 17 October 1978; 24 on 31 October 1978; 6 on 28 November 1978; and 2 on 17 December 1978.

One specimen (female).

SCOLOPACIDAE

Numenius minutus Gould **Little Whimbrel**

Moderately common visitor. Pair at airstrip on 30 October 1970; 20 over Mining Camp on 8 November 1970; 52 at airstrip from 20 October 1971 until early March 1972; 4 in open burnt area at lagoons 20 km south-west of Mining Camp on 30 October 1976; 1 at Mining Camp on 20 October 1978; flocks of 30, 500 and 50 at airstrip in October 1978; 8 at airstrip on 28 November 1978; and 8 on 3 January 1979.

One specimen (female).

Numenius phaeopus variegatus (Scopoli) **Whimbrel**

Common in tidal areas around Walsh Point and Lawley River delta. Usually ones, twos and threes; flock of 6 on 17 January 1973 and 1 of 21 on 2 July 1973 near Walsh Point. Mainly seen feeding along seaward edge of mangal.

Numenius madagascariensis (Linnaeus) **Eastern Curlew**

Recorded in 1970 (J.A.S.). Thirty-three on mudflats near Walsh Point on 2 July 1973; 10 (in ones and twos) along tidal creeks of Lawley River delta in October 1976.

Limosa lapponica baueri Naumann **Bar-tailed Godwit**

Twenty-two (3 in breeding plumage) on mudflats near Walsh Point on 4 March, 12 on 21 March and 2 on 14 April 1979 (J.A.S.).

Tringa nebularia (Gunnerus) **Greenshank**

One at Mining Camp on 7 December 1970; 1 by pool on plateau on 22 December 1971; a total of 4 near Walsh Point on 2 July 1973, 16 at Walsh Point on 21 February 1979, 7 on 21 March 1979 and 8 on 14 April 1979.

Tringa terek (Latham) **Terek Sandpiper**

Six near Walsh Point on 2 July 1973, 2 on 21 March 1979, and 10 in dead mangrove over water at mouth of Lawley River on 4 March 1979.

Ten specimens (2 female and 8 in spirit).

Tringa hypoleucos Linnaeus **Common Sandpiper**

One feeding among rocks at edge of mangal near Walsh Point on 17 January 1973, and 1 on a mudflat at Lawley River delta on 19 October 1976.

Tringa brevipes (Vieillot) **Grey-tailed Tattler**

One "tattler" [probably this species] near Walsh Point on 1 July 1973 (H.B.G.). Eight on mudflat near Walsh Point on 21 March 1979 and 2 on 14 April 1979 (J.A.S.).

Arenaria interpres interpres (Linnaeus) **Ruddy Turnstone**

One on bare island in Admiralty Gulf on 17 July 1970, and about 20 on mudflat near mouth of Lawley River on 17 October 1976.

Gallinago megala Swinhoe **Swinhoe's Snipe**

Only recorded at Airfield Swamp. About 20 flushed from *Eleocharis* and *Sorghum* growing in water 30 cm deep among *Melaleuca leucadendron.* Four in long grass (*Pseudopogonatherum contortum, Sorghum plumosum* and *Eriachne pauciflora*) at edge of Airfield Swamp on 31 January 1973.

Four specimens (1 male, 2 ?males and 1 female).

Calidris ruficollis (Pallas) **Red-necked Stint**

Two flocks of 50 at Walsh Point on 2 July 1973 and 4 on mudflat near Walsh Point on 4 March 1979.

Calidris ferruginea (Pontoppidan) **Curlew Sandpiper**

Seven, including one in nuptial plumage, at Walsh Point on 2 July 1973 (H.B.G.).

Limicola falcinellus sibriricus Dresser **Broad-billed Sandpiper**

Six on a mudflat at mouth of Lawley River on 17 October 1976.

RECURVIROSTRIDAE

Himantopus himantopus leucocephalus Gould **Black-winged Stilt**

One at pool near Mining Camp on 5 January 1973 (J.A.S.).

Recurvirostra novaehollandiae Vieillot **Red-necked Avocet**

Three on mudflat south of Walsh Point on 21 June 1975.

BURHINIDAE

Burhinus grallarius (Latham) **Bush Stone-curlew**

Common. Ones and twos, mainly on plateau. Seen feeding on tadpoles from drying pools. Nest on plateau on 6 October 1971 attended by both parents; they changed over on eggs about every 2 hours.

Esacus magnirostris (Vieillot) **Beach Stone-curlew**

Recorded between April and October 1970 (J.A.S.)

GLAREOLIDAE

Stiltia isabella (Vieillot) **Australian Pratincole**

Uncommon. Seen occasionally in open areas of plateau during dry season.

Glareola maldivarum Forster **Oriental Pratincole**

Common summer visitor. Recorded from November to January (mostly December). Single birds or in flocks of up to 1000. One flock of 500 at airstrip from 4 December 1970 until 16 January 1971. Usually seen after storms and cyclonic winds, often with Fork-tailed Swifts.

LARIDAE

Sterna nilotica Gmelin **Gull-billed Tern**

Five at Port Warrender on 2 July 1973 probably belonged to the Australian race *macrotarsa*. A specimen of the Asian race *S. nilotica affinis* was collected near the mouth of the Lawley River on 17 October 1976 (Johnstone, 1977). It was one of 4 feeding in shallow water at edge of mangal.

One specimen (unsexed).

Sterna bergii Lichtenstein **Crested Tern**

Seven near Walsh Point on 1 July 1973 (H.B.G.).

Sterna bengalensis Lesson **Lesser Crested Tern**

Moderately common in Admiralty Gulf. Mostly flocks of 2 to 20. A flock of 20 at Myres Island on 4 November 1976 contained some juveniles with speckled plumage and dull grey caps.

One specimen (unsexed).

Sterna hirundo longipennis Nordmann **Common Tern**

A flock of 3 and a flock of 6, feeding close to shore at Port Warrender on 4 November 1976. A flock of several hundred birds, possibly of this species, feeding at Port Warrender on 17 January 1973.

Two specimens (both male).

Sterna hybrida javanica Horsfield **Whiskered Tern**

Six feeding with White-winged Black Terns at Port Warrender on 4 November 1976. Five terns, probably of this species, at a small shallow pool at airstrip late February 1972; they remained for about 5 days.

Sterna leucoptera Temminck **White-winged Black Tern**

Three hawking over shallow water at Lawley River delta on 17 October 1976. A flock of 12 feeding close to shore with Whiskered and Common Terns on 4 November 1976.

Three specimens (1 male, 1 female and 1 unsexed).

Sterna sinensis Gmelin **Little Tern**

Twelve at Port Warrender on 2 July 1973. Several very small white terns, possibly this species, feeding with Common, White-winged Black and Whiskered Terns at Port Warrender on 4 November 1976.

COLUMBIDAE

Ducula bicolor spilorrhoa (Gray) **Torres Strait Pigeon**

Moderately common. Flocks (up to 6, occasionally up to 25). Favouring closed semi-deciduous vine forests with fruiting trees and shrubs including *Ficus* spp., *Mimusops elengi, Zizyphus quadrilocularis, Randia cochinchinensis, Celtis philippensis, Terminalia* spp. and *Eugenia* spp. Occasionally in shrubs among sandstones and in *Melaleuca leucadendron* along watercourses. In winter tending to disperse in small groups of 2 to 5 through adjacent woodlands, especially on plateau.

Thirteen specimens (8 male, 5 female).

Geopelia humeralis (Temminck) **Bar-shouldered Dove**

Common in closed semi-deciduous vine forests (including low closed forest) with *Ficus* and other fruiting trees, in dense thickets and scrubs in sandstones and along watercourses (*Eugenia, Ficus, Barringtonia, Pandanus* etc.) and in mangal. Nest with 2 eggs in vine thicket near Rail Creek on 21 October 1976. Nest with 2 half grown young at Mining Camp on 24 April 1979.

One specimen (male).

Geopelia striata placida Gould **Peaceful Dove**

Common to very common in woodlands on alluvial flats, around lagoons, at landward edge of mangal, and at Airfield Swamp, and in *Eucalyptus miniata, E. tetrodonta* woodland and palm savannah on the plateau. Moderately common in sandstone areas. Nest with 2 young on 27 March 1971.

One specimen (male).

Geopelia cuneata (Latham) **Diamond Dove**

Apparently only a dry season visitor, all 5 records falling between May and November. Largest flock 60, in October at open burnt area on plateau.

One specimen (male).

Chalcophaps indica longirostris Gould **Green-winged Pigeon**

Uncommon. In ones and twos. Favours the lowest stratum at edges of closed and low closed semi-deciduous vine forests, and thickets fringing watercourses.

Three specimens (1 male, 2 female).

Petrophassa albipennis Gould **White-quilled Rock Pigeon**

Common. Mainly ones, twos and small parties (up to 5). Rugged King Leopold Sandstones. Breeding season protracted: eggs in March, April, May, July and August; young in June and October.

Four specimens (3 male, 1 female).

Phaps chalcoptera (Latham) **Common Bronzewing**

Moderately common, in ones and twos, on plateau and alluvial flats. Nest with 2 eggs on 6 August 1971.

Geophaps smithii blaauwi Mathews **Partridge Pigeon**

Moderately common in flocks up to 20. Mainly among small sandstone boulders and soft grasses at junction of alluvial flats and sandstones. Also singly or in pairs on plateau during summer.

Eight specimens (3 male, 5 female).

PSITTACIDAE

Trichoglossus haematodus rubritorquis Vigors & Horsfield **Rainbow Lorikeet**

Uncommon to very common. Mainly twos and small flocks up to 20, occasionally larger flocks up to 100. Their distribution and abundance is dependent on availability of flowers. During summer 1973 they were largely confined to Airfield Swamp and major watercourses (*Melaleuca leucadendron* flowering). In winter 1975 they were dispersed through flowering *Eucalyptus tetrodonta, E. miniata* and flowering deciduous elements in vine forests such as *Bombax ceiba* and *Albizia lebbek*. Often seen in vine forests in October presumably sheltering from heat. Pair flew from hollow in eucalypt in October 1976.

Trichoglossus versicolor Lear **Varied Lorikeet**

Common to very common. Small flocks (up to 40). Usually feeding in flowering *Eucalyptus miniata, E. tetrodonta* and *Melaleuca leucadendron.*

Aprosmictus erythropterus (Gmelin) **Red-winged Parrot**

Common in flocks of 8 to 40 in semi-deciduous vine forests, feeding in fruiting trees. Moderately common in woodland on basalt and in creekside vegetation; scarce elsewhere. Seen feeding on fruits of *Cochlospermum fraseri.*

Platycercus venustus (Kuhl) **Northern Rosella**

Common. Mostly on the plateau in ones, twos, threes and fours. Woodland of *Eucalyptus tetrodonta, E. miniata* and *Livistona eastonii.* This species is more common here than elsewhere in Kimberley mainland (it is equally plentiful on Koolan Island).

One specimen (female).

Calyptorhynchus magnificus macrorhynchus Gould **Red-tailed Black Cockatoo**

Moderately common to very common. In flocks of less than 10 in June, July and August, "large numbers" in April (J.A.S.) and flocks of 400+ in September. The large flocks are probably dry season visitors. Seen feeding in *Eucalyptus miniata, Terminalia canescens* and *Grevillea refracta.*

One specimen (male).

Cacatua roseicapilla Vieillot **Galah**

One on plateau on 17 September 1978 and one 10 km north of Surveyors Pool on 3 December 1978 (J.A.S.).

Cacatua tenuirostris sanguinea Gould **Corella**

Uncommon. Flocks up to 15 at King Edward River crossing; on plateau on 19 June 1975 and two at Crystal Creek on 4 November 1976. One flew from hollow in tree at Mitchell Falls on 23 June 1975.

Cacatua galerita fitzroyi (Mathews) **Sulphur-crested Cockatoo**

Moderately common. Mainly semi-deciduous vine forests, fringing forest along major watercourses (especially cadjeputs) and thickets in sandstone gorges.

CUCULIDAE

Cuculus saturatus Blyth **Oriental Cuckoo**

Scarce to moderately common. Ones, twos and small flocks (up to 8). Observed in January, February, March and December. Mainly river-fringing forest, edge of mangal, Airfield Swamp and on plateau.

Cuculus pallidus (Latham) **Pallid Cuckoo**

Moderately common non-breeding visitor. Mostly in sandstone areas but also on plateau and alluvial flats. Six recorded on plateau between 23 and 25 November 1970.

Cuculus variolosus Vigors & Horsfield **Brush Cuckoo**

Moderately common. Mainly on plateau. Despite its immature plumage (pale lower mandible, orbital ring grey not yellow, rufous tinge to tail, wings and back, and barred underparts) a female collected at Camp Creek on 29 January 1973 had a hard-shelled egg in oviduct.

Three specimens (2 male, 1 female).

Chrysococcyx osculans (Gould) **Black-eared Cuckoo**

One in vine thicket at Lawley River on 7 January 1979 (J.A.S.).

Chrysococcyx basalis (Horsfield) **Horsfield's Bronze-cuckoo**

Scarce. Mainly on plateau.

Chrysococcyx malayanus minutillus Gould **Little Bronze-cuckoo**

Scarce in mangal. A juvenile being fed by 2 Large-billed Flyeaters (*Gerygone magnirostris*) was collected in dense mangal of *Avicennia marina* and *Aegiceras corniculatum* near mouth of Rail Creek on 19 October 1976. A cuckoo with a two-noted call heard in mangal near Walsh Point on 9 February 1973 was thought to be this species.

One specimen (immature, unsexed).

Eudynamys scolopacea subcyanocephala Mathews **Koel**

Moderately common summer visitor. Earliest record 17 October, latest 18 February. Confined to closed semi-deciduous vine forests and dense creekside vegetation. Often heard calling at night. Seen feeding on fruit of *Ficus virens* var. *sublanceolata*. Before attacking a Great Bowerbird a male Koel was observed in threat posture—wings lowered, tail fanned and neck outstretched.

Two specimens (male, wing 202 mm, wt 230 g, length 410 mm; female, wing 204 mm, wt 230 g, length 415 mm). These are considerably smaller than specimens in the Western Australian Museum of *E. s. cyanocephala* from south Kimberley and eastern Australia.

Scythrops novaehollandiae Latham **Channel-billed Cuckoo**

One flying south at Crystal Head on 25 January 1973.

Centropus phasianinus phasianinus (Latham) **Pheasant Coucal**

Moderately common. Mainly in tall rank grass and pandanus along watercourses and on alluvial flats; also edge of vine forests and along *Typha*-choked section of Rail Creek. Less common on plateau.

Nest with 4 eggs found on 26 March 1971, built in low shrub 0.6 m above ground among very dense 3 m high sorghum. Grass was curled around the well constructed nest, which was about the size of a football and lined with eucalypt leaves. Nest subsequently re-examined: 2 April, 1 egg hatched; 6 and 7 April 3 eggs hatched, the fourth egg missing; the young with black skin and white "hair". Young still present on 14 April but gone on 17 April.

Nest with 3 young found on 5 February 1979, built in tall grass 10 cm above ground. The young with black skin and white "hair", the oldest just starting to get feathers. The nest still contained egg shells but was otherwise clean and free of droppings. Nest with 4 feathered young and one addled egg found on 6 February 1979, built in dense grass 10 cm above ground, lined with eucalypt leaves and with a platform of grass approximately 20 cm long at entrance.

One specimen (male).

STRIGIDAE

Tyto alba delicatula (Gould) **Barn Owl**

Scarce. One seen on plateau on 17 January 1973 and 2 in semi-deciduous vine thickets near Rail Creek in October 1976.

One specimen (male).

Tyto novaehollandiae novaehollandiae (Stephens) **Masked Owl**

Scarce. One in dense *Melaleuca* on 23 July 1972, 1 in a *Pandanus* at edge of mangal near Rail Creek on 18 October 1976. The latter bird was still using the roost on 21 October.

Ninox rufa rufa (Gould) **Rufous Owl**

Scarce. Two sightings in the same patch of semi-deciduous vine thicket; one on 20 January, the other on 1 February 1973.

Ninox connivens connivens (Latham) **Barking Owl**

One in a eucalypt on plateau on 17 January 1973 and one calling at Crystal Creek on 2 November 1976.

Ninox novaeseelandiae boobook (Latham) **Boobook Owl**

The most common owl in area; seen or heard in all habitats. Adult with a juvenile just able to fly at Lawley River on 24 November 1978.

One specimen (male).

PODARGIDAE

Podargus strigoides (Latham) **Tawny Frogmouth**

Moderately common. Mainly on plateau and in eucalypt woodland on basalt (including edge of mangal). Nest with 2 eggs near Mitchell Falls on 22 November 1970. Nest with 2 fresh eggs near mouth of Lawley River on 24 November 1978.

One specimen (male).

AEGOTHELIDAE

Aegotheles cristatus leucogaster Gould **Australian Owlet-nightjar**

Moderately common. Mainly on plateau. Also 1 in burnt area at Crystal Creek, and 1 calling regularly from hollow of *Avicennia marina* in mangal at Rail Creek.

Two specimens (1 male, 1 female).

CAPRIMULGIDAE

Eurostopodus guttatus (Vigors & Horsfield) **Spotted Nightjar**

Moderately common. Mainly open areas on the plateau (roads, airstrip and exposed basalt flats etc.). One in sandstones at Crystal Creek.

Two specimens (male).

APODIDAE

Hirundapus caudacutus caudacatus (Latham) **Spine-tailed Swift**

Forty-two over Mining Camp on 22 December 1972 (J.A.S.).

Apus pacificus pacificus (Latham) **Fork-tailed Swift**

Moderately common. Mainly a summer visitor. Flocks of 50 to several hundreds (often with Oriental Pratincoles), usually associated with thunderstorms. Earliest record 28 September, latest 2 April.

ALCEDINIDAE

Alcedo azurea ruficollaris (Bankier) **Azure Kingfisher**

Moderately common. Mainly along small watercourses with overhanging vegetation: pools fringed with *Pandanus* and *Melaleuca*, and tidal creeks with mangroves.

One specimen (female).

Dacelo leachii leachii Vigors & Horsfield **Blue-winged Kookaburra**

Moderately common. Palm savannah on plateau, fringing vegetation along watercourses (mainly *Melaleuca leucadendron*), eucalypt woodland on sandstones (*E. tetrodonta* and *E. miniata*), and open woodland on basalts. Nest with 4 eggs in tree near Airfield Swamp on 12 October 1971; 4 fully grown young left this nest on 10 November 1971.

Halcyon pyrrhopygia Gould **Red-backed Kingfisher**

Uncommon. Mostly in open woodland on plateau and basalt. Scarce in sandstone areas. Probably mainly a dry season visitor.

One specimen (unsexed).

Halcyon sancta sancta Vigors & Horsfield **Sacred Kingfisher**

Scarce resident and uncommon winter visitor. Mainly in mangal, creekside vegetation, edge of semi-deciduous vine forests and occasionally woodland. Slight increase in numbers during dry season (especially in mangal) when resident population is augmented by southern visitors.

One specimen (male).

MEROPIDAE

Merops ornatus Latham **Rainbow Bee-eater**

Common. Pairs and small parties (up to 20). In all habitats, but mainly open areas on plateau and open basalt flats. Mainly a winter visitor, earliest flock 18 March, latest flock 21 November. A freshly dug nest found at Crusher on 13 October 1978, and 2 adults carrying food to a nest tunnel at Lawley River on 24 November 1978.

CORACIIDAE

Eurystomus orientalis pacificus (Latham) **Dollarbird**

Moderately common summer visitor. Mainly ones and twos. Edge of semi-deciduous vine forests (usually perched in tall *Bombax* or *Albizia*), creekside vegetation with tall *Melaleuca*, mangal, and occasionally woodland on plateau. Earliest records 4 October 1970, 26 September 1971, 15 October 1972, 10 September 1974 and 15 October 1976.

PITTIDAE

Pitta iris Gould **Rainbow Pitta**

Moderately common. Only in more luxuriant blocks of semi-deciduous vine forest, where they frequent the forest floor and spiny shrub layer. Never seen flying higher than 2 m. Adults more often identified by their territorial call "I lost my whip", by the distinctive sound of their hopping, or by the sound of their raking of litter for insects and snails.

A nest constructed of sticks and lined with fine grass was found on the ground on 3 February 1973. It resembled a collapsed Great Bowerbird bower and was 30 cm across and 20 cm high; the chamber was 12.5 cm across. The 5 eggs were pale white with brown blotches slightly more numerous on the larger end. An old nest made of sticks and dead leaves was 0.6 m up in a fork of a small shrub, and another old nest with one addled egg was found against base of huge *Bombax*.

Nine specimens (5 male, 4 female).

ALAUDIDAE

Mirafra javanica Horsfield **Horsfield's Bushlark**

Moderately common. Mainly open areas: airstrip, cleared area around crushing plant, edge of Airfield Swamp and treeless basalt and alluvial flats. Seen displaying in January 1973, and heard calling at night over airstrip.

Eight nests, 6 with 3 eggs, 1 with 2 eggs and 1 with 1 egg at airstrip on 23 and 24 March 1979, and 4 nests, 3 with 3 eggs and 1 with hatchlings also at airstrip on 26 March 1979. All 12 nests were checked on 27 March; 11 contained young and the nest with 1 egg on the 23 March contained 3 eggs. Most nests were built beneath small tussocks of grass and were dome shaped with a side entrance at ground level.

Three specimens (2 male, 1 female).

HIRUNDINIDAE

Hirundo rustica gutturalis Scopoli **Barn Swallow**

Two swallows flying over crushing plant on 30 October 1976 may have been this species.

Hirundo nigricans nigricans Vieillot **Tree Martin**

Scarce winter visitor. Ones, twos and small flocks up to 20. Mainly over plateau and along coast. No records after 20 October.

One specimen (? female).

Hirundo ariel (Gould) **Fairy Martin**

Four birds, probably this species, circling cliffs at Mitchell Falls on 30 October 1976.

MOTACILLIDAE

Anthus novaeseelandiae australis Vieillot **Richard's Pipit**

One at airstrip from 9 to 16 February 1979 (J.A.S.).

CAMPEPHAGIDAE

Coracina novaehollandiae novaehollandiae (Gmelin) **Black-faced Cuckoo-shrike**

Common on plateau, where often seen in small parties up to 8. Elsewhere scarce to moderately common in ones and twos. All wooded habitats except mangal. Attracted to *Ficus* and other fruiting trees.

Coracina papuensis hypoleuca (Gould) **Little Cuckoo-shrike**

Moderately common. Mostly ones and twos in river-fringing forest and at Airfield Swamp. Also in edge of vine forests and in open woodland. Attracted to stands of *Melaleuca leucadendron*, and flowering *Brachychiton* sp., *Eucalyptus miniata, and E. tetrodonta.*

A nest, made mostly of spider web, 7 m up on a horizontal fork of a bloodwood 2 km east of Mitchell Falls contained 2 heavily incubated eggs on 1 November 1976.

Three specimens (1 male, 1 female and 1 in spirit).

Coracina tenuirostris tenuirostris (Jardine) **Cicadabird**

Scarce (5 seen). Only recorded in semi-deciduous vine forests. Observed feeding in canopy in *Albizia lebbek, Bombax ceiba* and *Ficus* spp. Gut of 1 specimen contained a large red ant.

Three specimens (2 males, 1 female).

Lalage sueurii tricolor (Swainson) **White-winged Triller**

Moderately common. Mostly ones, twos and threes, occasionally small flocks (up to 12) and once a migratory flock of 20 including several adult males. Open woodland with flowering trees and shrubs, mainly *Eucalyptus miniata, E. tetrodonta, Cochlospermum fraseri* and *Grevillea* spp.

One specimen (male).

Lalage leucomela rufiventris (Gray) **Varied Triller**

Moderately common. Ones and twos. Mainly semi-deciduous vine forests. Also thickets of *Ficus, Terminalia, Melaleuca* and *Pandanus* along watercourses, and evergreen thickets in sandstone areas. Attracted to fruiting trees (especially *Ficus*) and flowering eucalypts.

Six specimens (3 male, 3 female).

PACHYCEPHALIDAE

Microeca tormenti Mathews **Brown-tailed Flycatcher**

Moderately common. Confined to mangal. Observed in tall *Avicennia, Bruguiera, Aegiceras* and *Ceriops* at Rail Creek and in mangal south of Walsh Point.

Two specimens (1 male, 1 female).

Microeca leucophaea leucophaea (Latham) **Jacky Winter**

Uncommon to moderately common. Mainly open eucalypt woodland on plateau, basalt and alluvial flats. Nest with 2 eggs on plateau on 12 October 1978 and a flimsy nest with 2 eggs in a *Terminalia* near Mining Camp on 2 November 1978.

Two specimens (1 male, 1 female).

Eopsaltria pulverulenta (Bonaparte) **Mangrove Robin**

Confined to mangal. Common at Walsh Point, moderately common at Rail Creek and Crystal Creek. Mainly dense forests of *Rhizophora stylosa, Aegiceras corniculatum* and *Avicennia marina*; occasionally thickets of *Ceriops* and *Bruguiera.*

Five specimens (2 male, 3 female).

Poecilodryas superciliosa cerviniventris (Gould) **White-browed Robin**

Uncommon. Mostly in pairs (once 3 together). Dense thickets and scrubs of *Ficus, Melaleuca, Terminalia, Pandanus* and *Barringtonia* along watercourses. Pair in open semi-deciduous vine forest near Mitchell Falls on 14 November 1971 building a nest on 21 November. The call is a parrot-like whistle "te te doo".

Four specimens (3 male, 1 unsexed).

Pachycephala caledonica melanura Gould **Mangrove Golden Whistler**

Common in mangal. Favouring thickets of *Ceriops tagal* and *Aegiceras corniculatum.*

Females from Rail Creek have more yellow on underparts than females from Walsh Point. However Rail Creek specimens are still not as yellow as females from Wyndham and localities further east. The area between Wyndham and the Lawley River delta is thus a zone of intergradation between yellow-bellied (*spinicauda*) and plain-bellied (*melanura*) females. The subspecific name *hilli* has been applied to these intergrades.

Fourteen specimens (7 male, 6 female, 1 unsexed).

Pachycephala rufiventris rufiventris (Latham) **Rufous Whistler**

Moderately common in all wooded country, especially dense vegetation along watercourses and at edge of vine forests; also *Eucalyptus miniata, E. tetrodonta* woodland on plateau.

Six specimens (3 male, 3 female).

Colluricincla megarhyncha parvula Gould **Little Shrike-thrush**

Common in semi-deciduous vine thickets and forests, and in evergreen scrubs in sandstone areas. Frequenting the lower storey of vine forest in trees and shrubs including *Ficus, Terminalia* and *Exocarpus* and the lowest spiny shrub layer of *Zizyphus, Abrus, Capparis* and *Flagellaria.* Moderately common in mangal.

Fifteen specimens (8 male, 6 female, 1 unsexed).

Colluricincla woodwardi Hartert **Sandstone Shrike-thrush**

Moderately common. Sandstone areas with cliffs and jumbled boulders. A new nest made entirely of spinifex rootlets was found in a rock crevice under a sandstone shelter on 14 January 1973. A nest with 3 fresh eggs in sandstone near Mitchell Falls on 22 October 1978, and another with 3 eggs in sandstone at Camp Creek on 7 November 1978.

Colluricincla harmonica harmonica (Latham) **Grey Shrike-thrush**

Uncommon. Single birds in open woodland. Mostly on plateau but also sandstone areas and edge of mangal.

MONARCHIDAE

Rhipidura rufifrons dryas Gould **Rufous Fantail**

One male collected by W.H.B. in semi-deciduous vine forest along edge of Camp Creek on 1 August 1973.

Rhipidura fuliginosa phasiana DeVis **Grey Fantail**

Moderately common in mangal. Observed feeding in *Rhizophora, Avicennia, Aegiceras, Bruguiera* and *Ceriops.*

Five specimens (2 male, 1 female, 2 unsexed).

Rhipidura leucophrys leucophrys (Latham) **Willie Wagtail**

Uncommon resident and moderately common winter visitor. Mainly open areas on plateau: around Airfield Swamp, Mining Camp and the bauxite crushing plant. Occasionally along watercourses in sandstones and on alluvial flats.

Rhipidura rufiventris isura Gould **Northern Fantail**

Common in semi-deciduous vine forest; river-fringing forests and thickets; open evergreen thickets at foot of sandstone cliffs and edge of mangal. Moderately common on plateau at Airfield Swamp and in *Eucalyptus miniata, E. tetrodonta* woodland.

Two half-finished nests, both made of spider web, were found at Rail Creek in October 1976. One was on a thin horizontal branch of a small shrub overhanging a creek and the other in a *Ficus* overhanging a cliff.

Four specimens (2 male, 1 female, 1 ?female).

Myiagra ruficollis mimikae Ogilvie-Grant **Broad-billed Flycatcher**

Moderately common. Confined to mangal, mainly tall stands of *Avicennia marina* and thickets of *Aegiceras corniculatum.*

Five specimens (2 male, 2 female, 1 unsexed).

Myiagra rubecula concinna Gould **Leaden Flycatcher**

Moderately common in semi-deciduous vine forests and dense vegetation along watercourses (*Melaleuca, Ficus, Terminalia, Pandanus* etc.). Uncommon in scrubs and thickets at foot of cliffs, open woodland, and mangal.

Nine specimens (6 male, 3 female).

Myiagra inquieta nana (Gould) **Restless Flycatcher**

Moderately common. Mainly dense *Melaleuca* and *Pandanus* along watercourses; also edge of vine forests and vine scrubs. Nest with 3 eggs in a *Melaleuca* over water at Crystal Head on 17 October 1971. Nest with 2 eggs near airstrip on 27 October 1978, built 3 m up in tree and decorated with lichens. Fresh empty nest 5 m up in tree on 28 October 1978.

Myiagra alecto rufolateralis (Gray) **Shining Flycatcher**

Uncommon. Confined to mangroves, where it favours dense forests and thickets of *Rhizophora stylosa, Aegiceras corniculatum* and *Avicennia marina.*

Three specimens (1 male, 2 female).

ORTHONYCHIDAE

Pomatostomus temporalis rubeculus (Gould) **Grey-crowned Babbler**

Generally scarce. Open woodland in palm savannah on plateau; occasionally woodlands on basalts and sandstones. Mostly small parties (up to 8). Nest with 2 eggs near airstrip on 27 October 1978 and nest with 3 eggs on 5 November 1978.

Two specimens (1 male and 1 in spirit).

ACANTHIZIDAE

Gerygone olivacea rogersi Mathews **White-throated Flyeater**

Uncommon. Mainly in *Melaleuca leucadendron* at Airfield Swamp and along watercourses. Also *Acacia-Eucalyptus* woodland on plateau and basalt. Attracted to flowering *Melaleuca* and *Eucalyptus.*

Two specimens (males).

Gerygone levigaster levigaster Gould **Mangrove Flyeater**

One collected in mangrove-*Melaleuca* interface about 10 km south-south-east of Walsh Point in February 1979 (W.H.B.).

Gerygone magnirostris magnirostris Gould **Large-billed Flyeater**

Common to moderately common. Confined to mangal where it usually feeds in canopies of *Avicennia marina, Aegiceras corniculatum* and *Rhizophora stylosa.* Pair feeding an immature Little Bronze Cuckoo at Rail Creek on 19 October 1976.

Six specimens (4 male, 2 female).

Gerygone chloronota chloronota Gould **Green-backed Flyeater**

Moderately common in all strata of semi-deciduous vine forests and thickets. Also evergreen thickets in sandstone areas and dense vegetation of *Ficus, Melaleuca, Terminalia, Barringtonia* etc. along watercourses.

Nine specimens (5 male, 4 female).

Smicrornis brevirostris (Gould) **Weebill**

Moderately common in small parties (up to 10). Mainly in eucalypts on plateau but also in eucalypts on alluvial flats and in *Melaleuca leucadendron* at Airfield Swamp.

One specimen (female).

MALURIDAE

Amytornis housei (Milligan) **Black Grasswren**

Moderately common, in pairs and small parties (up to 6). Heavily dissected sandstone areas to west of plateau with dense spinifex (*Triodia* and *Plectrachne*), *Acacia, Xanthostemon* and scattered *Eucalyptus* and *Ficus.* Not seen in King Edward River drainage.

Breeding during wet season. A female with a soft-shelled egg in oviduct was collected at Crystal Head on 25 January 1973, and an empty fresh nest was found here. Nest dimensions 20 cm long, 15 cm wide, entrance hole about centre of nest. It was placed in a large spinifex clump in the open, made of spinifex and lined with soft grass. A fledgling just able to fly was caught by hand on 11 February 1973.

Twelve specimens (4 male, 7 female, 1 unsexed).

Malurus lamberti rogersi Mathews **Variegated Fairy-wren**

Moderately common in pairs and small parties (up to 5). In both heavily dissected sandstones to west of plateau and the more gentle sandstones along King Edward River. Mainly areas with spinifex (*Triodia* and *Plectrachne*), *Acacia, Xanthostemon, Calytrix, Owenia* and *Ficus.*

A nest with 3 eggs, built in low spinifex 25 cm above ground in sandstone edge of Lawley River on 10 December 1978.

Two specimens (males).

Malurus melanocephalus (Latham) **Red-backed Fairy-wren**

Common in pairs and small parties. Mainly in bunch grasses on plateau and basalts; also among *Typha* and *Pandanus* along watercourses and at edge of mangal.

SYLVIIDAE

Megalurus timoriensis Wallace **Tawny Grassbird**

Scarce. Several in tall bunch grasses immediately behind mangal on Rail Creek. These very secretive birds shared this habitat with much more common Golden-headed Fantail Warblers and Red-backed Wrens. One collected at Airfield Swamp, in grassland edge of *Melaleuca* on 28 September 1978.

Two specimens (males).

Cincloramphus mathewsi Iredale **Rufous Songlark**

Uncommon. Single birds and groups (up to 4). Probably a dry-season visitor. Only on plateau where it was seen in open grassy areas (Airfield Swamp and Mining Camp) in June and November.

One specimen (female).

Cincloramphus cruralis (Vigors & Horsfield) **Brown Songlark**

One at edge of airstrip on 20 June 1975.

Cisticola exilis (Vigors & Horsfield) **Golden-headed Fantail Warbler**

Very common in tall bunch grasses immediately behind mangal on Rail Creek. Common along watercourses in grass, *Pandanus* and *Typha*, at edge of vine forests and thickets, in woodland on basalt, and at Airfield Swamp.

Nest with 3 hatchlings, built 10 cm above ground in tall grass at Camp Creek on 9 February 1979. Nest dome shaped about size of tennis ball, with side entrance, and made of dried fibres of spear grass and spider web.

Ten specimens (6 male, 3 female and 1 in spirit).

DAPHOENOSITTIDAE

Daphoenositta chrysoptera leucoptera (Gould) **Australian Sittella**

Moderately common on plateau and alluvial flats in small flocks (up to 5), elsewhere scarce. Mainly open eucalypt woodland of *E. miniata* and *E. tetrodonta* on plateau and bloodwoods and *Melaleuca* along watercourses.

Two specimens (female).

CLIMACTERIDAE

Climacteris melanura melanura Gould **Black-tailed Tree-creeper**

Scarce (only 8 seen). Mainly in extreme west and south of study area. Usually in *Eucalyptus tectifica* woodland.

DICAEIDAE

Dicaeum hirundinaceum hirundinaceum (Shaw) **Mistletoe-bird**

Uncommon to moderately common. Mainly semi-deciduous vine forests and thickets, dense vegetation along watercourses, evergreen thickets in sandstone areas, and mangal.

PARDALOTIDAE

Pardalotus striatus uropygialis Gould **Striated Pardalote**

Very common in breeding season; hundreds nesting in grader spoil at edge of road on plateau in June and July. Moderately common in open eucalypt woodland in off-breeding season. Construction of roads and grid lines has provided the Striated Pardalote with many more nest sites than were available naturally (banks of creeks and rivers etc.). The small number of resident birds on the plateau indicates that after breeding, individuals disperse over a very large area.

One specimen (male).

ZOSTEROPIDAE

Zosterops lutea Gould **Yellow White-eye**

Moderately common, in flocks of 6-20. Mangal, and scrub along creeks running into mangal.

MELIPHAGIDAE

Lichmera indistincta indistincta (Vigors & Horsfield) **Brown Honeyeater**

Common to very common. In all wooded habitats. Seen feeding at flowering mangroves (*Aegiceras corniculatum* and *Ceriops tagal*), *Eucalyptus miniata, E. tetrodonta, Melaleuca leucadendron, M. argentea, Grevillea pteridifolia, Albizia lebbek* and *Brachychiton* sp.

Myzomela erythrocephala erythrocephala Gould **Red-headed Honeyeater**

Common. Mainly in mangal, but also scrub at adjacent creek margins. A fledgling just able to fly fed by adults in mangal at Rail Creek on 20 October 1976.

Two specimens (1 male, 1 female).

Meliphaga albilineata (White) **White-lined Honeyeater**

Moderately common. Rugged sandstone areas west and east of plateau. Favouring deep gorges and gullies. Seen feeding in *Ficus* spp., *Melaleuca* spp., *Xanthostemon paradoxus, Eucalyptus miniata* and *Acacia* spp. A nest with 2 eggs, built of dried grass and spinifex and woven with spiders web suspended in small shrub under large sandstone shelter at Crystal Creek on 11 November 1978. Eggs pink with chestnut band formed by spots on larger end.

Three specimens (2 male, 1 female).

Meliphaga virescens (Vieillot) **Singing Honeyeater**

Scarce. One in open woodland near King Edward River crossing on 3 July 1973, and 1 in open woodland near Rail Creek on 19 October 1976.

Meliphaga plumula (Gould) **Yellow-fronted Honeyeater**

Scarce, only 3 records (2 on plateau and 1 at King Edward River crossing). Probably only a rare visitor; this species favours drier parts of the Kimberley.

Meliphaga flavescens flavescens (Gould) **Yellow-tinted Honeyeater**

Moderately common in tall trees (*Melaleuca* and *Eucalyptus*) along watercourses. Also at edge of semi-deciduous vine forests, when trees and shrubs such as *Albizia* are flowering.

Meliphaga unicolor (Gould) **White-gaped Honeyeater**

Moderately common. Mainly semi-deciduous vine forests and thickets, edge of mangal, and creekside vegetation. Observed feeding on trunks of *Ceriops tagal* in mangal and at all levels of vine forests.

Four specimens (3 male, 1 female).

Melithreptus albogularis Gould **White-throated Honeyeater**

Scarce to moderately common. Mainly well-wooded areas along creeks and rivers and edge of vine forests. Seen feeding in flowering *Livistona eastonii, Albizia lebbek, Eucalyptus miniata* and *Melaleuca* spp.

Three specimens (2 male, 1 female).

Melithreptus gularis laetior Gould **Black-chinned Honeyeater**

Scarce. Woodlands with flowering eucalypts, cadjeputs and *Grevillea* on plateau and alluvial flats.

Melithreptus cyanotis albipennis (Gould) **Blue-faced Honeyeater**

Scarce resident and moderately common dry-season visitor. Pairs and small parties (up to 5, once flock of 20) on plateau and at edge of vine forests and thickets and in creekside vegetation. Also feeding in flowering *Melaleuca leucadendron* at Airfield Swamp.

One specimen (female).

Philemon citreogularis citreogularis (Gould) **Little Friarbird**

Scarce to moderately common. Woodland habitats especially *Eucalyptus miniata*, bloodwoods, *Melaleuca* spp. and edge of mangal.

Two specimens (1 male, 1 female).

Philemon argenticeps argenticeps (Gould) **Silver-crowned Friarbird**

Moderately common to common. Favouring semi-deciduous vine forests and thickets with flowering and fruiting trees especially *Ficus* spp., *Bombax ceiba, Albizia lebbek, Terminalia* and *Eugenia.* Also open eucalypt woodland of *Eucalyptus miniata* and *E. tetrodonta*, creekside vegetation in sandstone areas and flowering *Melaleuca leucadendron* at Airfield Swamp.

Nest with young on 3 March 1972; nest with 2 eggs in eucalypt on 13 January 1973; and nest with 2 eggs in patch of vine scrub on 17 October 1976.

Three specimens (1 male, 2 female).

Cissomela pectoralis (Gould) **Banded Honeyeater**

Scarce to common. Mainly small flocks (up to 10). Visiting flowering *Eucalyptus miniata, E. tetrodonta, Melaleuca leucadendron, M. argentea* and other flowering trees and shrubs along watercourses.

Ramsayornis fasciatus (Gould) **Bar-breasted Honeyeater**

Moderately common in small flocks (up to 6). Vegetation along watercourses and at Airfield Swamp especially *Melaleuca leucadendron.* Also feeding in flowering eucalypts at edge of vine forests and scrubs and in flowering *Gardenia* in sandstones.

A nest with 2 eggs, built entirely of paperbark and spider web at Lawley River on 10 December 1978, and another nest with 2 eggs and 1 young at Mitchell Falls on 8 April 1979.

Four specimens (2 male, 2 female).

Conopophila rufogularis (Gould) **Rufous-throated Honeyeater**

Scarce. Several seen in open *Eucalyptus miniata* woodland with *Grevillea* at King Edward River crossing; one in open woodland edge of plateau and one at Mitchell River Station.

Manorina flavigula (Gould) **Yellow-throated Miner**

Moderately common in pairs and small parties (up to 5). Mostly on plateau but also in woodland on sandstone and along watercourses.

PLOCEIDAE

Neochmia phaeton phaeton (Hombron & Jacquinot) **Crimson Finch**

Common. Mostly small parties (up to 12), in pandanus and tall rank grass along watercourses. Breeding from early December to late April. Clutch size 3 to 5, nests often constructed but not used.

Poephila bichenovii annulosa (Gould) **Double-barred Finch**

Common to very common. Mainly small flocks (up to 20) but several flocks of up to 50 recorded. Open grassy areas especially along watercourses; also edges of mangal and vine forests. Often forming mixed flocks with Long-tailed Finches. Breeding from January to August, clutch size 3 to 4.

Poephila personata personata Gould **Masked Finch**

One record (J.A.S.): pair feeding young on plateau on 27 July 1970.

Poephila acuticauda (Gould) **Long-tailed Finch**

Scarce to common. Usually small flocks up to 10 (occasionally 20-30). Mainly grassy areas along watercourses; one flock of 25 in open burnt area at edge of plateau. Adult carrying nesting material in open woodland on 7 November 1976.

Erythrura gouldiae (Gould) **Gouldian Finch**

Moderately common to common. Mainly a dry-season visitor. Small flocks (up to 20) during wet season and large flocks (up to 200) in July, August and September (1972, 1973 and 1978). Usually on plateau and alluvial flats. Occasionally accompanying Long-tailed Finches. A pair feeding young on plateau on 27 July 1970.

Lonchura castaneothorax castaneothorax (Gould) **Chestnut-breasted Mannikin**

Common. Usually small flocks (up to 30) but often large flocks (up to 200). Mainly around Mining Camp feeding on lawns. Also soft grasses around Airfield Swamp and on alluvial flats. At Rail Creek a flock of 100 on 18 October 1976 in dense *Aegiceras corniculatum* beside an area of soft grass, and a flock of 200 on 19 October 1976 in dense *Typha*.

Two nests, one with 2 eggs the other with 3 eggs at Lawley River on 5 March 1979. Both nests were woven in tall grass 1.5 m above the ground.

Four specimens (2 male, 2 female).

Lonchura pectoralis (Gould) **Pictorella Mannikin**

An unconfirmed report by J.A.S. on the plateau on 8 June 1970.

ORIOLIDAE

Oriolus sagittatus (Latham) **Olive-backed Oriole**

Moderately common in semi-deciduous vine forests and thickets with fruiting trees and shrubs including *Ficus* spp., *Eugenia* and *Terminalia*. Less frequent in thickets along watercourses and in flowering eucalypts. Occasionally feeding with Figbirds.

Two specimens (1 female, 1 in spirit).

Oriolus flavocinctus flavocinctus (King) **Yellow Oriole**

Moderately common, mostly ones and twos. Semi-deciduous vine forests and scrubs, mangal, dense vegetation along watercourses and woodland. Often feeding in fruiting trees of *Ficus, Eugenia, Celtis* and *Terminalia*; also seen eating Green Tree-ants *Oecophylla smaragdina*.

Six specimens (5 males, 1 female).

Sphecotheres viridis flaviventris Gould **Figbird**

Uncommon to moderately common. Mostly small parties (up to 8). Semi-deciduous vine forests and thickets with fruiting trees especially *Ficus, Eugenia* and *Ganophyllum*; also tall dense *Eucalyptus miniata* and *Terminalia* fringing forest at Mertens Falls.

Seven specimens (4 male, 3 female).

DICRURIDAE

Dicrurus megarhynchus bracteatus Gould **Spangled Drongo**

Scarce. A pair in patch of closed forest by Mitchell River on 14 November 1971; one near mouth of Lawley River on 12 November 1978; one in vine thicket on plateau on 26 November 1978; and one 15 km north of King Edward River crossing on 7 December 1978 (J.A.S.).

GRALLINIDAE

Grallina cyanoleuca (Latham) **Magpie-lark**

Scarce resident and common winter visitor. Ones and twos during wet season along watercourses, around lagoons and Airfield Swamp, and at Mining Camp. Winter visitors arrive in April, May and June and leave in October. Flocks of up to 50 during dry season in open areas such as Mining Camp, Airfield Swamp, airfield and basalt and alluvial flats. On 5 April 1972 J.A.S. recorded birds beginning to arrive and several days later numbers had increased to over 50 around Mining Camp; after a thunderstorm on 22 October 1972 the flock had diminished to 12 birds.

ARTAMIDAE

Artamus leucorhynchus (Linnaeus) **White-breasted Wood-swallow**

Scarce or uncommon. Around large pools on major watercourses fringed with *Melaleuca.* Also lagoons, edge of mangal, and alluvial flats close to coast.

Artamus personatus (Gould) **Masked Wood-swallow**

Scarce. Only recorded in south-eastern parts of study area. A flock of 50 at the King Edward River crossing on 3 July 1973 and flock of 10 at 10 km west of crossing on 24 June 1975. Attracted to flowering *Grevillea pteridifolia.*

Artamus cinereus melanops Gould **Black-faced Wood-swallow**

Scarce. Pairs and small parties (up to 12). Open woodland on basalt, alluvial flats and parts of plateau. Breeding on 19 October 1971; nest on top of a *Livistona* palm.

Artamus minor Vieillot **Little Wood-swallow**

Moderately common. Pairs and small groups (up to 6). Mainly open areas on plateau (e.g. Airfield Swamp, lagoons, bauxite crushing plant and Mining Camp) and coastal basalts.

CRACTICIDAE

Cracticus torquatus latens Ford **Grey Butcherbird**

Moderately common in open vine thickets at foot of basalt cliffs and along watercourses near Rail Creek. Uncommon along *Melaleuca*-lined creeks in sandstone areas and also edge of vine forests on plateau.

One specimen (male).

Cracticus nigrogularis (Gould) **Pied Butcherbird**

Very common on plateau, moderately common elsewhere.

One specimen (male).

PARADISAEIDAE

Ptilonorhynchus nuchalis nuchalis Jardine & Selby **Great Bowerbird**

Common in closed forest with fruiting trees and shrubs, especially *Ficus, Randia, Terminalia, Eugenia* and *Celtis.* Bowers often built in lower tangled layer of prickly *Capparis, Abrus* and *Flagellaria indica.* Moderately common in open vine scrubs at foot of cliffs and along watercourses; also in mangal.

CORVIDAE

Corvus orru salvadorii Finsch **Australian Crow**

Scarce, in ones and twos. Attracted to fires and habitation. This is probably the resident *Corvus* in the area. However the Little Crow *C. bennetti* may visit the area during the dry season, e.g. a flock of 40+ crows seen by J.A.S. on 26 May 1971 was probably *C. bennetti.*

One specimen (in spirit).

DISCUSSION

A total of 219 species (137 non-passerine and 82 passerine) have been recorded from the area covered in this paper. Few additional passerine species could be expected; however, more work along the coast would no doubt add many visiting waders to this list.

The avifauna is typical of the subhumid north-west sector of the Kimberley, and has much in common with the Prince Regent River Reserve to the south-west (Storr *et al.*, 1975) and Napier Broome Bay to the north-east (Hill, 1911). That the number of species in the Mitchell Plateau area is greater than at Prince Regent River Reserve and Napier Broome Bay is attributed to the wider variety of habitats in general, and the better-developed closed forests in particular.

The semi-deciduous vine forests are the richest on the Kimberley mainland (lush forests also occur on the Osborn Islands). They are best developed on basalts around the edge of the plateau. These forests and thickets contain many fruiting trees and shrubs and have deep leaf litter rich in snails and insects. Birds confined to the vine forests include the Scrub Fowl (*Megapodius reinwardt*), Rufous Owl (*Ninox rufa*), Rainbow Pitta (*Pitta iris*) and Cicadabird (*Coracina tenuirostris*). Other species which feed in these forests include the Torres Strait Pigeon (*Ducula bicolor*), Bar-shouldered Dove (*Geopelia humeralis*), Green-winged Pigeon (*Chalcophaps indica*), Olive-backed Oriole (*Oriolus sagittatus*), Yellow Oriole (*O. falvocinctus*), Figbird (*Sphecotheres viridis*), Koel (*Eudynamys scolopacea*), Varied Triller (*Lalage leucomela*), Little Shrike-thrush (*Colluricincla megarhyncha*) and Great Bowerbird (*Ptilonorhynchus nuchalis*).

No single block of forest is generally large enough to contain all of these species. Moreover the amount of fruit available at any one place fluctuates widely so that most of the birds which are wholly or mainly frugivorous must move over a large area in search of food. This movement is probably instrumental in the establishment of new forests by the deposition of seeds and in carrying seeds from one block of forest to another.

The coastal thickets and scrubs tend to support fewer birds, especially the ground dwelling and canopy feeding species. The flowering and fruiting trees in these more open thickets are utilised by the Koel, Little Shrike-thrush, Green-backed Flyeater (*Gerygone chloronota*), Olive-backed Oriole, Yellow Oriole, Silver-crowned Friar-bird (*Philemon argenticeps*), and the Great Bowerbird.

The riverine scrubs and thickets are inhabited by the rare White-browed Robin (*Poecilodryas superciliosa*) and Rufous Fantail (*Rhipidura rufifrons*) and other species such as the Northern Fantail (*Rhipidura rufiventris*) and Leaden Flycatcher (*Myiagra rubecula*).

Alluvial flats and open woodlands are habitats which are common throughout the Kimberley and contain many wide-ranging species such as the Brown Quail (*Coturnix ypsilophora*), Partridge Pigeon (*Geophaps smithii*) and the Red-backed Fairy-wren (*Malurus melanocephalus*) on the alluvial flats and the Rufous Whistler (*Pachycephala rufiventris*), Jacky Winter (*Microeca leucophaea*) and Little Friarbird (*Philemon citreogularis*) in the open woodlands.

The absence of domestic stock is an important factor affecting the number of bird species in the riverine vegetation, alluvial flats and open woodlands. Elsewhere in the Kimberley there has been extensive destruction, especially to riverine vegetation and alluvial flats. This has led to the decline of the White-browed Robin (Smith & Johnstone, 1977) and the Partridge Pigeon (Johnstone, in press.).

Lateritic plateau with dense eucalypts and palms is scarce or absent elsewhere in the Kimberley. Although no species is confined to the plateau it supports good populations of the Emu (*Dromaius novaehollandiae*), Bustard (*Otis australis*), Northern Rosella (*Platycercus venustus*), Striated Pardalote (*Pardalotus striatus*) and Pied Butcherbird (*Cracticus nigrogularis*).

The rugged King Leopold Sandstones with their deep hummock grasses support a full complement of sandstone birds: White-quilled Rock Pigeon (*Petrophassa albipennis*), Variegated Fairy-wren (*Malurus lamberti*), Sandstone Shrike-thrush (*Colluricincla woodwardi*), Black Grasswren (*Amytornis housei*) and White-lined Honeyeater (*Meliphaga albilineata*). All are found on the flatter sandstones to the east, with the exception of the Black Grasswren which is not found north-east of Surveyors Pool and Crystal Creek and is endemic to the wet north-west sector of the Kimberley.

The extensive blocks of mangal in Admiralty Gulf contain 14 species of mangrove birds. There is one interesting absentee, the White-breasted Whistler (*Pachycephala lanioides*); another, the Mangrove Flyeater (*Gerygone levigaster*) is rarely found in the mangal along the subhumid north-west coast. The White-breasted Whistler has a large break in its range between the Prince Regent River and Cambridge Gulf. The Little Shrike-thrush, a bird of similar size and bill shape to that of the Whistler, occurs in the mangal within this area and may have replaced the Whistler through competitive exclusion. The Mangrove Flyeater has also been collected in the mangal at Prince Regent River and Napier Broome Bay. In south-western Kimberley and at Cambridge Gulf this species prefers stands of *Melaleuca* usually near or contiguous with the mangal. At Admiralty Gulf the mangal is mostly backed by open woodland and in some areas by vine thickets, both of which are generally unsuitable for this flyeater. Hill in 1909-10 collected 5 specimens of the Mangrove Flyeater at Napier Broome Bay, but we did not record it there in June 1975; there, instead, we found the Large-billed Flyeater (*Gerygone magnirostris*) to be the resident flyeater. The distribution, taxonomy and ecology of Western Australian mangrove birds will be dealt with in a later paper by one of us (R.E.J.).

Hill recorded and collected birds at Napier Broome Bay from August 1909 to July 1910. He observed 8 species (excluding waders and oceanic seabirds) not seen in our study area: Coot (*Fulica atra*), Red-crowned Pigeon (*Ptilinopus regina*), Silver Gull (*Larus novaehollandiae*), Red Goshawk (*Accipiter radiatus*), Hooded Robin (*Petroica cucullata*), Crested Shrike-tit (*Falcunculus frontatus*), Crested Bellbird

(*Oreoica gutturalis*) and Black Honeyeater (*Certhionyx nigra*). The Coot and Black Honeyeater would be vagrants there. The Red-crowned Pigeon has been recorded on the nearby Osborn Islands and may eventually be found in the coastal thickets of Admiralty Gulf. The Silver Gull, uncommon along most of the Kimberley coast, would no doubt have been attracted to the mission station which was very close to the beach. The Crested Bellbird and Hooded Robin are absent from the subhumid zone, preferring the drier and less densely wooded parts of the Kimberley. The Red Goshawk and the Crested Shrike-tit are rare in the Kimberley with only 3 records of the former and 1 of the latter (Storr, 1980).

Despite the regularity of the climate the following birds show considerable nomadism: the two lorikeets and many of the honeyeaters, including the Black-chinned (*Melithreptus gularis*), Yellow-fronted (*Meliphaga plumula*) and Banded (*Cissomela pectoralis*), wander over most of the Kimberley in search of flowering trees and shrubs.

The migrant species, which made up 13.5% of the bird fauna, can be divided into 5 main groups:

1. Non-breeding summer visitors from the Palaearctic, e.g. Mongolian Sand Plover (*Charadrius mongolus*), Little Whimbrel (*Numenius minutus*), Oriental Pratincole (*Glareola maldivarum*), Swinhoe's Snipe (*Gallinago megala*), Asian Gull-billed Tern (*Sterna nilotica affinis*), Oriental Cuckoo (*Cuculus saturatus*), Fork-tailed Swift (*Apus pacificus*) and Barn Swallow (*Hirundo rustica*).
2. Breeding summer visitors from the north, e.g. Koel (*Eudynamys scolopacea*) and Dollarbird (*Eurystomus orientalis*).
3. Non-breeding summer visitors from the south, e.g. Marsh Harrier (*Circus aeruginosus*) and Pallid Cuckoo (*Cuculus pallidus*).
4. Passage migrants, e.g. Channel-billed Cuckoo (*Scythrops novaehollandiae*) and many of the waders such as the Greenshank (*Tringa nebularia*).
5. Dry season visitors. The Tree Martin (*Hirundo nigricans*) and the White-winged Triller (*Lalage sueurii*) are probably winter migrants from the south. Others are mainly visitors from drier parts of the Kimberley, e.g. Red-tailed Black Cockatoo (*Calyptorhynchus magnificus*), Diamond Dove (*Geopelia cuneata*), Rufous Songlark (*Cincloramphus mathewsi*), Blue-faced Honeyeater (*Melithreptus cyanotis*) and Little Crow (*Corvus bennetti*).

Populations of some resident species, Rainbow Bee-eater (*Merops ornatus*), Sacred Kingfisher (*Halycon sancta*) and the Gouldian Finch (*Erythrura gouldiae*), are greatly augmented by winter visitors.

REFERENCES

BEARD, J.S. (1976)—The monsoon forests of the Admiralty Gulf, Western Australia. *Vegetatio* **31:** 177-192.

HILL, G.F. (1911)—Field notes on the birds of Kimberley, north-west Australia. *Emu* **10:** 258-290.

JOHNSTONE, R.E. (1977)—An Asian Gull-billed Tern in Western Australia. *West. Aust. Nat.* **14:** 26-27.

JOHNSTONE, R.E. Notes on the distribution, ecology and taxonomy of the Partridge Pigeon (*Geophaps smithii*) and Spinifex Pigeon (*Geophaps plumifera*) in Western Australia. (In press.)

MAYR, E. (1943)—Notes on Australian birds (11). *Emu* **43:** 3-17.

SMITH, L.A. & JOHNSTONE, R.E. (1977)—Status of the Purple-crowned Wren (*Malurus coronatus*) and Buff-sided Robin (*Poecilodryas superciliosa*) in Western Australia. *West. Aust. Nat.* **13:** 185-188.

STORR, G.M. (1980)—Birds of the Kimberley Division, Western Australia. Spec. Publs West. Aust. Mus. No.11.

STORR, G.M., JOHNSTONE, R.E., DELL, J. & SMITH, L.A. (1975)—Birds of the Prince Regent River Reserve, north-west Kimberley, Western Australia. *In* A biological survey of the Prince Regent River Reserve, north-west Kimberley, Western Australia in August, 1974. *Wildl. Res. Bull. West. Aust.* **3:** 75-84.

Received 6 February 1979   Accepted 23 October 1979   Published 28 February 1981

PART 6

AMPHIBIANS AND REPTILES OF MITCHELL PLATEAU AND ADJACENT COAST AND LOWLANDS, KIMBERLEY, WESTERN AUSTRALIA

L.A. SMITH*

and

R.E. JOHNSTONE*

INTRODUCTION

The area covered by this paper occupies 4400 km² between 14°25′ and 15°08′S and 125°40′ and 126°12′E. It is bounded by the Mitchell River in the west and south, Admiralty Gulf in the north and the upper King Edward River in the east. The area comprises a lateritic plateau with basalt slopes and surrounding areas of rugged sandstone, alluvial flats and varied coastal habitats.

Little was known of the herpetofauna of Mitchell Plateau area before 1970 (Wilson, Part 1, this publication). In that year naturalist J.A. Smith commenced work with Amax Exploration Ltd (now Alumax). It was mainly through his interest and efforts that we were able to spend part of the 1972-73 wet season in the area.

This paper is based on collections and observations made during visits to the study area by us from 14 January-24 February 1973, 18-24 June 1975 and 15 October-7 November 1976. We also include specimens collected by A.K. Lee of Monash University (June 1965), J.A. Smith (1970-73), D.J. Kitchener of the Western Australian Museum (September 1971), W.H. Butler (21 July-10 August 1973) and M.J. Tyler of Adelaide University (January-February 1978).

Wilson (Part 1, this publication) briefly discusses the climate and geology of the area. The vegetation has been studied by Beard (1976) and Hnatiuk & Kenneally (Part 2, this publication).

We consider that, as in other parts of Western Australia, the nature of the substrate rather than the vegetation is the major factor determining the distribution of reptiles in the Kimberley (Storr 1964).

For the purposes of this paper we have divided substrates in the area into three types: laterite (**Figs 2 & 3**, Johnstone & Smith, Part 5, this publication), basalt, including alluvial flats (**Figs 6 & 7**, Part 5, this publication) and King Leopold Sandstones (**Fig. 9**, Part 5, this publication). The various collecting areas on these substrate types are prefixed L, B and S respectively.

---

*Department of Ornithology & Herpetology, Western Australian Museum, Francis Street, Perth, Western Australia. 6000.

METHODS

Most collecting was done at 9 sites **(Fig. 1)**.

B1 and **B2** (Rail Creek in 14°41′S, 125°52′E; Walsh Point, Port Warrender, in 14°34′S, 125°50′E respectively). These comprise woodlands on exposed basalt scree slopes with areas of open semi-deciduous vine thickets at the base of basalt cliffs. There are extensive mangals at both sites.

L1 (within 8 km radius of Mining Camp, centred in 14°50′S, 125°50′E). Collecting here was mostly confined to 4 areas: the immediate vicinity of Mining Camp; the bauxite crushing plant, semi-deciduous vine forest 8 km south of Mining Camp, and Airfield Swamp.

L2 (Glauert Lagoon in 14°57′S, 125°57′E). Collecting here was confined to the alluvial soils and riverine vegetation (mostly *Pandanus*).

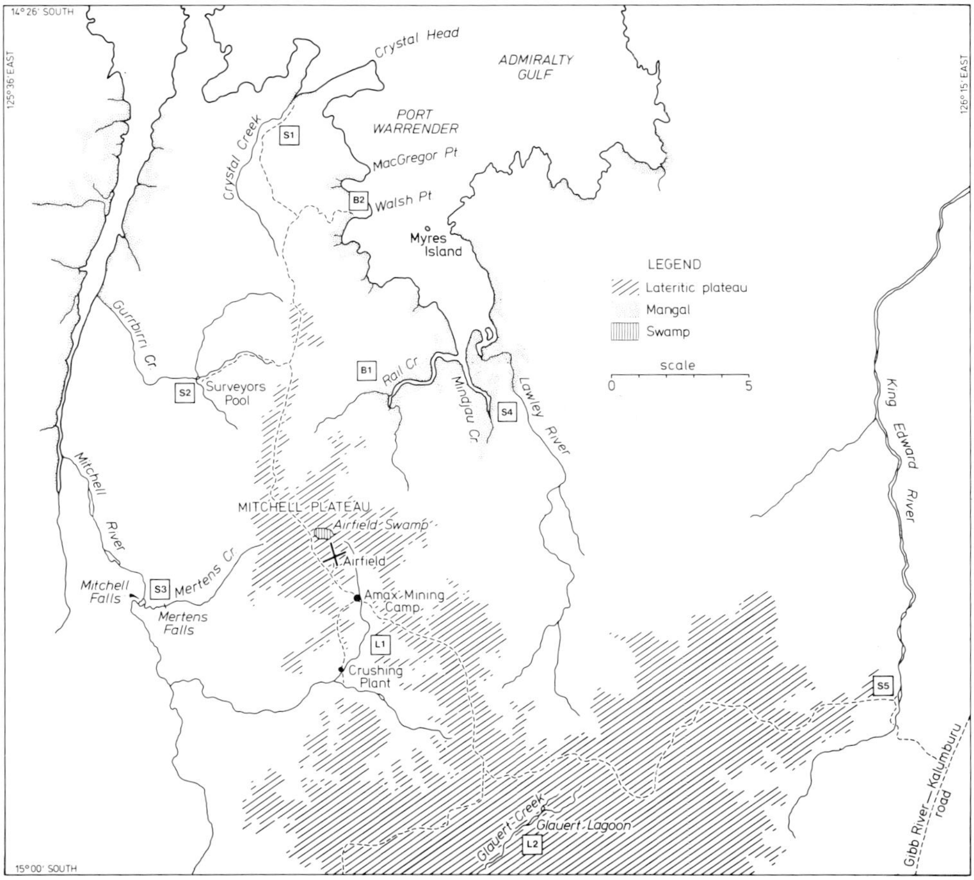

Fig. 1: Map of the Mitchell Plateau region, Western Australia showing main collecting areas.

S1 and **S2** (Crystal Creek in 14°30′S, 125°47′E and Surveyors Pool in 14°41′S, 125°44′E [base camp 3] respectively). These comprise King Leopold Sandstones and woodlands on alluvial soils.

S3, S4 and **S5** (Mitchell River Falls in 14°50′S, 125°42′E; Lawley River in 14°57′S 126°00′E; King Edward River in 14°52′S, 126°12′E respectively). These comprise King Leopold Sandstones.

Localities of the few specimens not collected in these areas are given in detail.

All but a few of the specimens (namely those collected by M.J. Tyler *et al.* in 1978 and mostly housed in the South Australian Museum) are lodged in the Western Australian Museum under registered numbers R28182-227, 43002-43391, 43508-43537, 44258-44275, 44279-44326, R45038-50, 56162-56321, 56337-56432, 57156-57160, 57195-57211, R58300-332, 59775-86 and 59951-59971.

We seldom assess relative abundance but some indication of it can be deduced from the number of specimens collected. Where no number follows a locality, the record is based on observation only.

ANNOTATED LIST OF SPECIES

ANURA

FAMILY LEPTODACTYLIDAE **Ground Frogs**

Cyclorana australis (Gray)

Recorded from L1 (14) and S1 (1).

From Crystal Creek from burning *Plectrachne pungens*; specimens from Mining Camp from long grass.

Cyclorana longipes Tyler & Martin

Recorded at L1 (15).

Limnodynastes convexiusculus (Macleay)

Recorded at L1 (9).

Six from Mining Camp and 3 from long grass and damp soil at Airfield Swamp.

Limnodynastes ornatus (Gray)

Recorded from L1 (26).

All in grass around Mining Camp.

Ranidella bilingua (Martin, Tyler & Davies)

Recorded from B1 (5), L1 (21) and L2 (16).

In mud and short grass at edge of watercourses.

Uperoleia spp.

Recorded at L1 (39).

M.J. Tyler, University of Adelaide, informs us (*in litt.*, 26.5.78) that there are two undescribed species of *Uperoleia* from the Mitchell Plateau.

FAMILY HYLIDAE **Tree Frogs**

Litoria bicolor (Gray)

Recorded at L1 (48), L2 (4).

Twenty-six from Airfield Swamp from leaves of *Melaleuca leucadendron* and *Pseudopogonatherum contortum*, 4 from Glauert Lagoon were in *Pandanus*, most of remainder (from around Mining Camp) on stems and leaves of long grass.

Litoria caerulea (White)

Recorded at L1 (28).

Almost all from in and around buildings at Mining Camp.

Litoria cavernicola Tyler & Davies

Recorded at S2 (3).

From deep within a cave in sandstone.

Litoria coplandi (Tyler)

Recorded at B1 (6), L1 (12) and S1 (35).

In rocky pools in Rail Creek and in grass on banks of Camp Creek.

Litoria latopalmata Günther

Recorded at L1 (30) and S1 (1).

Mostly from edges of pools around Mining Camp. The specimen from Crystal Creek was active among jumbled sandstone at night.

Litoria meiriana (Tyler)

Recorded at B1 (6) and 15 km N of Mining Camp (2).

Specimens at Rail Creek from in and around a small pool in basalt.

Litoria microbelos (Cogger)

Recorded from L1 (29) and L2 (1).

In short grass at edge of Camp Creek.

Litoria nasuta (Gray)

Recorded at L1 (35) and L2 (5).

Mostly on ground in short grass.

Litoria rothii (De Vis)

Recorded at B1 (4), L1 (22) and S1 (3).

Mostly in shrubs and low trees or in buildings. Those from Crystal Creek collected at night while active among jumbled sandstone.

Litoria rubella (Gray)

Recorded at B1 (2), L1 (25) and S1 (2).

In pools on basalt at Rail and Camp Creeks. From spinifex on laterite and sandstone at L1 and S1 respectively.

Litoria wotjulumensis (Copland)

Recorded at B1 (12), L1 (9) and L2 (2).

Specimens at Rail Creek from around small pool on basalt.

REPTILIA

FAMILY CHELUIDAE **Side-necked Turtles**

Chelodina rugosa Ogilby

Recorded at L1 (1) and S3 (1).

One in *Eucalyptus miniata, E. tetrodonta, Livistona* woodland about 3 km from Camp Creek; the other in a pool in sandstone.

FAMILY GEKKONIDAE **Geckos**

Crenadactylus ocellatus naso Storr

Recorded from S1 (8) and S4 (3).

All from burning spinifex during day.

Diplodactylus ciliaris Boulenger

Recorded at L1 (2) and S5 (1).

Both specimens from Mining Camp, high (up to 5 m) in eucalypts.

Diplodactylus mcmillani Storr

Recorded at S1 (26), S3 (2), S4 (11) and S5 (1).

All but King Edward River specimen from burning spinifex on sandstone during day.

Diplodactylus stenodactylus Boulenger

Recorded at S1 (1) and S5 (1).

Crystal Creek specimen from small bare area of laterite at night.

Gehyra australis Gray

Recorded at L1 (2) and S1 (3).

Two on walls of buildings at Mining Camp and 3 among jumbled sandstone or on trees. All at night.

Gehyra nana Storr

Recorded at L1 (2), S1 (12), S2 (1) and S5 (2).

Nine from burning spinifex during day; remainder among jumbled sandstone boulders at night.

Gehyra xenopus Storr

Recorded at S1 (16), S2 (3) and S5 (3).

All but 1 active on massive sandstone blocks or on trees at night. One from Mitchell River Falls in cave during day. This was the gecko most commonly seen among the sandstone at night.

Heteronotia binoei (Gray)

Recorded at L1 (6), S2 (1) and S5 (1).

Under lateritic boulders during day. Active among sandstone boulders at night.

Heteronotia spelea (Kluge)

Recorded at S2 (1).

Active among sandstone boulders at night.

Nephrurus asper Günther

Recorded at L1 (1) and S5 (1).

One under lateritic boulder during day and the other on flat sandstone at night.

Oedura marmorata Gray

Recorded at S1 (3) and S2 (2).

Active among sandstone boulders and in shrubs at night.

Oedura rhombifera Gray

Recorded at B1 (1) and S1 (9).

Active in shrubs and among sandstone boulders at night.

Pseudothecadactylus lindneri cavaticus Cogger

Recorded at S1 (4), S2 (2) and S3 (1).

Three from caves during day. Remainder collected at night in *Ficus* spp. growing from walls of sandstone chasm. Seen leaping up to 2 m from one branch to another or dropping 6 m to ground when pursued.

Rhynchoedura ornata Günther

Recorded at L1 (1).

From stomach of *Lialis burtonis* burnt from spinifex on laterite.

FAMILY PYGOPODIDAE — **Legless Lizards**

Delma borea Kluge

Recorded at B2 (1), L1 (6) and S1 (7).

From burning spinifex on sandstone and spinifex on laterite. Also among lateritic and basalt boulders.

Lialis burtonis Gray

Recorded at B1 (1), L1 (5) and S4 (1).

Mostly from burning spinifex on laterite (once from spinifex on sandstone). One specimen's stomach contained a *Rhynchoedura ornata.*

Pygopus nigriceps (Fischer)

Recorded at L1 (1).

Active on bare laterite at night.

FAMILY AGAMIDAE — **Dragon Lizards**

Amphibolurus microlepidotus Glauert

Recorded at S1 (2).

One among sandstone and spinifex during day; the other 1.5 m up in a wattle at night.

Chelosania brunnea Gray

Recorded at L1 (1).

Chlamydosaurus kingii Gray

Recorded at L1, B2 (1) and on alluvial flats east of Surveyors Pool (2).

Diporiphora albilabris albilabris Storr

Recorded at L1 (16), S1 (2) and S5 (3).

All but 5 on laterite.

Diporiphora bennettii bennettii (Gray)

Recorded at B2 (10), L1 (17) S1 (8), S2 (5) and S5 (3).

Mostly on laterite and basalt.

Diporiphora convergens Storr

Recorded from S1 (1).

Diporiphora magna Storr

Recorded at S1 (1), S5 (3).

All on sandstone.

Diporiphora superba Storr

Recorded at S2 (5).

The holotype of this species was collected from a wattle tree at the edge of Mertens Creek. One of the other specimens ran across a large sandstone slab and leapt into a small fig tree about 1 m high. When alive it was a brilliant lime green all over except for a bright yellow oblique dash 1 cm long behind the axilla.

Lophognathus gilberti gilberti Gray

Recorded at L1 (6) and S3 (1).

In trees and leaf litter at margins of creeks.

FAMILY SCINCIDAE **Skinks**

Carlia amax Storr

Recorded from B2 (1) and L1 (12).

All on laterite and basalt (mostly laterite).

Carlia foliorum (De Vis)

Recorded at L1 (6), L2 (1), S1 (1), S2 (1) and 8 km north-east of Mitchell River Station (1).

All but one (from burning spinifex on sandstone) were from bunch grasses on alluvial soils or basalts.

Carlia gracilis Storr

Recorded from L1 (1), L2 (13) and S1 (1).

Specimens from edge of Glauert Lagoon were active in soft mud among *Nymphaea* and *Eleocharis* spp. Colour description of live males: orange-red dorso-lateral stripe from axilla of arm to axilla of leg. Throat and upper labials iridescent blue-green. Remainder of back and tail brownish-black with pearly blue opalescence.

Carlia johnstonei Storr

Recorded from B1 (22), B2 (15), L1 (16), S1 (1) and S2 (1).

One from leaf litter under boab tree in sandstone; otherwise specimens from basalts and laterite (68.5% of these from leaf litter in semi-deciduous vine forests and thickets).

Carlia triacantha (Mitchell)

Recorded from B1 (1), L1 (4), S1 (17), S2 (7) and S5 (11).

Eighty-three per cent of specimens from sandstone areas (mostly burnt from *Triodia* and *Plectrachne*, but also in leaf litter).

Cryptoblepharus megastictus Storr

Recorded at B1 (2) and L1 (2).

All on basalt: on boulders in semi-deciduous vine thickets or margins of creeks.

Cryptoblepharus plagiocephalus (Cocteau)

Recorded at B2 (1), L1 (7), S1 (4) and S2 (1).

All but 2, one of which was on sandstone, the other on laterite, were on the trunks of trees.

Ctenotus decaneurus Storr

Recorded from L1 (5).

One from burning spinifex on laterite, and another from under slab of basalt.

Ctenotus inornatus (Gray)

Recorded at L1 (7), S1 (54), S2 (4), and S5 (5).

Mostly from burning *Plectrachne* and *Triodia* on sandstone but also from burning *Triodia* on laterite.

Ctenotus mastigura Storr

Recorded at L1 (3).

All among spinifex on laterite.

Ctenotus robustus Storr

Recorded at L1 (2) and S2 (2).

Two burnt from spinifex on laterite and 2 among *Pandanus* and long grass at edge of watercourse on alluvial flat.

Menetia maini Storr

Recorded at B2 (1).

On deep red basaltic soil with Sorghum and sparse eucalypts.

Morethia ruficauda ruficauda (Lucas & Frost)

Recorded at B1 (1), S1 (1) and S3.

Two among spinifex on sandstone and 1 on basalt in semi-deciduous vine thicket with *Carlia johnstonei.*

Notoscincus ornatus wotjulum (Glauert)

Recorded at L1 (2) and S1 (3).

One under tussock of grass on alluvial flat and another in leaf litter under boab tree on sandstone.

Proablepharus tenuis (Broom)

Recorded from S1 (11).

All from burning spinifex on sandstone.

Sphenomorphus brongersmai Storr

Recorded at L1 (3) and S1 (1).

One specimen in breakback trap along creek in *Melaleuca leucadendron, Pandanus* and grasses. Another among *Livistona*, eucalypts and grasses was caught 30 minutes after dark.

The *Sphenomorphus* sp. listed for Prince Regent Reserve by Storr & Smith (1975) is *S. brongersmai* (*fide* A.E. Greer, pers. comm.).

Sphenomorphus isolepis (Boulenger)

Recorded at B2 (2), L1 (7), S1 (1), S2 (2), S3 (1) and 8 km north-east of Mitchell River Station (1).

From burning *Plectrachne* on sandstone and *Triodia* on laterite and were seen in leaf litter in semi-deciduous vine thickets during day. Active at night on laterite with bunch grasses and among sandstone with spinifex.

Tiliqua scincoides (Shaw)

Recorded from L1 (2).

FAMILY VARANIDAE **Monitors**

Varanus acanthurus Boulenger

Recorded at B2 (1), L1 (2), S1 (1) and 12 km north of Mining Camp (1).

Varanus glauerti Mertens

Recorded at B1 (1), B2 (1), S1 (1) and S2 (1).

One from burning *Plectrachne pungens* on sandstone, and another taken from a narrow crack in a basalt boulder along a watercourse.

Varanus glebopalma Mitchell

Recorded at S1 (1).

Active among spinifex and jumbled sandstone.

Varanus gouldii (Gray)

Collected at L1 (1).

In *Eucalyptus miniata, E. tetrodonta* and *Livistona* woodland.

Varanus mertensi Glauert

Recorded at S2 (1) and S3.

In pools along creeks.

Varanus timorensis scalaris Mertens

Recorded at L1 (1) and S2 (1).

BOIDAE **Pythons**

Aspidites melanocephalus (Krefft)

J.A. Smith (pers. comm.) says this species was encountered on several occasions during his 3-year stay in the area. One specimen, run over 16 km south of Walsh Point, had a total length of 3023 mm.

Liasis gilberti (Gray).

Recorded at S1 (2) and S2 (1).

The 2 specimens from Crystal Creek were from burning *Plectrachne pungens* on sandstone. This little known species is common in the Kimberley.

Liasis olivaceus Gray

Recorded at S1 (1).

Shot in *Melaleuca leucadendron* on Crystal Creek. Gut contained a Torres Strait Pigeon (*Ducula bicolor*).

Python sp.

Recorded at S3 (1).

From sandstone overhang.

This is the only known specimen of this undescribed species. It is being described L.A. Smith (in prep.).

Python spilotus variegatus (Gray)

Recorded from L1.

J.A. Smith collected this specimen 8 km north of Mining Camp on 6 June 1971. It was photographed and released. Copy of photograph in Western Australian Museum (R 60607).

FAMILY COLUBRIDAE **Rear-fanged and Fangless Snakes**

Boiga fusca (Gray)

Recorded from L1 (1) and S1 (1).

Dendrelaphis punctulatus (Gray)

Recorded at B2 (1) and L1 (1).

One active in *Eucalyptus miniata, E. tetrodonta, Livistona* woodland; the other active in canopy of mangal.

Natrix mairii (Gray)

Recorded at L1 (1).

In Camp Creek at Mining Camp.

FAMILY ELAPIDAE **Front-fanged Snakes**

Demansia olivacea olivacea (Gray)

One specimen collected between Mitchell and Lawley Rivers in 1967.

Demansia papuensis melaena Storr

Recorded from L1 (1).

Demansia simplex Storr

Recorded at L1 (2).

Both on basalt along watercourses with *Terminalia* and *Pandanus.*

Denisonia pallidiceps (Günther)

Recorded from L1 (1) and S5 (1).

Furina christieana (Fry)

Recorded at B1 (1).

Active in long grass on basalt at night.

Pseudechis australis (Gray)

Recorded at S2 (1).

Fairly common (J.A. Smith, pers. comm.). The largest measured specimen had a total length of 2516 mm.

FAMILY HYDROPHIIDAE **Sea Snakes**

Hydrelaps darwiniensis Boulenger

Recorded from B2 (1).

From mangal just south of Walsh Point.

FAMILY CROCODYLIDAE **Crocodiles**

Crocodylus johnstoni Krefft

Recorded upstream as far as Mitchell River Station on Mitchell River, Mining Camp on Camp Creek and Surveyors Pool on Gurrbirri Creek, usually singly.

Crocodylus porosus Schneider

See Messel *et al.* (1977) for information on past and present numbers and size classes in the Lawley and Mitchell River systems.

DISCUSSION

A total of 1018 specimens representing 86 species were collected in the study area during all seasons except autumn.

All but 3 families (Cheluidae, Varanidae and Typhlopidae) must be completely or nearly completely listed. Five species recorded from the Prince Regent River Reserve (Storr & Smith 1975), viz. *Emydura australis, Lerista walkeri, Omolepida maxima, Varanus mitchelli* and *Acanthophis antarcticus,* and *Denisonia punctata* (collected just outside the study area in 15°33′S, 126°09′E) were not collected on the Mitchell Plateau surveys but are expected to occur there.

The herpetofauna of the Mitchell Plateau area, like the Prince Regent River Reservesis rich in taxa of the subhumid zone of north-west Kimberley. The number of predominately diurnal species (agamids, skinks, and varanids) is similar in both areas; however, there are wide differences in the number of cryptozoic species recorded, particularly the geckos and snakes. The additional 19 species in the Mitchell Plateau area probably reflects the greater number of man-hours spent there over 3 seasons, rather than any real difference in the herpetofauna of the 2 areas.

A number of species endemic to north-west Kimberley are found in the King Leopold Sandstones around Mitchell Plateau. They include *Litoria cavernicola,*

Gehyra xenopus, Crenadactylus ocellatus naso, Diplodactylus mcmillani, Pseudothecadactylus lindneri cavaticus, D. convergens, D. superba, Amphibolurus microlepidotus and an undescribed species of *Python*. These species are restricted to sandstone areas. The other species that were only collected among sandstone in the study area, occur in various habitats elsewhere in the Kimberley.

Only 1 species (*Cryptoblepharus megastictus*) appears to be restricted to exposed basalt, at least in the Mitchell Plateau area.

Although 9 species (*Rhynchoedura ornata, Pygopus nigriceps, Chelosania brunnea, Ctenotus mastigura, Tiliqua scincoides, Varanus gouldii, Python spilotus, Natrix mairii* and *Demansia papuensis melaena*) were found only on the lateritic plateau, only 1 of these (*Ctenotus mastigura*) has not been found in other habitats in other parts of the Kimberley.

The Mitchell Plateau area is the type locality of 18 taxa, viz. *Cyclorana longipes* (Tyler & Martin, 1977), *Litoria cavernicola* (Tyler & Davies, 1979), *Uperoleia* spp. nov. (2 species Tyler *et al.*, in prep.), *Crenadactylus ocellatus naso* (Storr, 1978), *Diplodactylus mcmillani* (Storr, 1978), *Gehyra nana* (Storr, 1978), *G. xenopus* (Storr, 1978), *Pseudothecadactylus lindneri cavaticus* (Cogger, 1975), *Diporiphora albilabris albilabris* (Storr, 1974a), *D. convergens* (Storr, 1974a), *D. superba* (Storr, 1974a), *Carlia amax* (Storr, 1974b), *C. gracilis* (Storr, 1974b), *C. johnstonei* (Storr, 1974b), *Cryptoblepharus magastictus* (Storr, 1976), *Ctenotus mastigura* (Storr, 1975) and *Python* sp. nov. (Smith, in prep.).

ACKNOWLEDGEMENTS

We are indebted to Amax Exploration for their assistance and hospitality during our three visits to Mitchell Plateau.

We would like to thank J.A. Smith (formerly of Amax Exploration) for data and specimens collected during his stay at Mitchell Plateau and K.F. Kenneally, Western Australian Herbarium, for identifying the plants. G.M. Storr, Head, Department of Ornithology and Herpetology, Western Australian Museum, read the manuscript.

REFERENCES

BEARD, J.S. (1976)—The monsoon forests of the Admiralty Gulf, Western Australia. *Vegetatio* **31:** 177-192.

BUTLER, W.H. (1979)—First Record of the Taipan for Western Australia. *West. Aust. Nat.* **14:** 134.

COGGER, H.G. (1975)—New lizards of the genus *Pseudothecadactylus* (Lacertilia: Gekkonidae) from Arnhem Land and northwest Australia. *Rec. Aust. Mus.* **30:** 87-97.

MESSEL, H., BURBIDGE, A.A., WELLS, A.G. & GREEN, W.J. (1977)—The status of the salt-water crocodile in some river systems of the north-west Kimberley, Western Australia. *Dep. Fish. Wildl. West. Aust.* no. 24.

STORR, G.M. (1964)—Some aspects of the geography of Australian reptiles. *Senck. Biol.* **45:** 577-589.

STORR, G.M. (1974a)—Agamid lizards of the genera *Caimanops, Physignathus* and *Diporiphora* in Western Australia and the Northern Territory. *Rec. West. Aust. Mus.* **3:** 121-146.

STORR, G.M. (1974b)—The genus *Carlia* (Lacertilia, Scincidae) in Western Australia and the Northern Territory. *Rec. West. Aust. Mus.* **3:** 151-165.

STORR, G.M. (1976)—The genus *Cryptoblepharus* (Lacertilia, Scincidae) in Western Australia. *Rec. West. Aust. Mus.* **4:** 53-63.

STORR, G.M. (1978)—Seven new gekkonid lizards from Western Australia. *Rec. West. Aust. Mus.* **6:** 337-352.

STORR, G.M. & SMITH, L.A. (1975)—Amphibians and reptiles of the Prince Regent River Reserve, north-western Australia. *In*: A biological survey of the Prince Regent River Reserve, north-west Kimberley, Western Australia, in August 1974. (Eds Miles, J.M. & Burbidge, A.A.) *Wildl. Res. Bull. West. Aust.* No. 3: 75-84.

TYLER, M.J. & DAVIES, M. (1979)—A new species of cave-dwelling hylid frog from Mitchell Plateau, Western Australia. *Trans. R. Soc. S. Aust.* **103:** 149-153.

TYLER, M.J. & MARTIN, A.A. (1977)—Taxonomic studies in some leptodactylid frogs of the genus *Cyclorana* Steindachner. *Rec. S. Aust. Mus.* **17:** 261-276.

ADDENDA

W.H. Butler has made 3 more visits to the study area since this paper was submitted. During his visits of 25 September-8 October and 26 October-7 November 1978 and 21 February-6 March 1979 he collected the following additional species for the area: a single large *Oedura* (probably a new species), *Lerista walkeri, Varanus tristis tristis, Vermicella annulata multifasciata* and *Oxyuranus scutellatus* (see Butler 1979 for details of *O. scutellatuss*).

Received 6 February 1979 Accepted 23 October 1979 Published 28 February 1981

PART 7

FRESHWATER FISH FAUNA OF THE MITCHELL PLATEAU AREA, KIMBERLEY, WESTERN AUSTRALIA

J.B. HUTCHINS*

INTRODUCTION

Prior to this survey, the freshwater fish fauna of the Mitchell Plateau area was unknown. Mees and Douglas had made a small collection of brackish and freshwater fishes from the nearby King Edward River at Kalumburu in 1960 (Mees 1962). Collecting trips by Allen in 1974 and Hutchins in 1975, both of the Western Australian Museum, were made, respectively, to the Prince Regent and Roe Rivers to the west of the plateau and to the Drysdale and Carson Rivers to the east (Allen 1975; Hutchins 1977).

This report deals with fishes collected from fresh water during the November 1976 survey. A future paper by this author will deal with the species collectèd from marine and brackish water areas of the Mitchell Plateau.

METHODS

Selection of collecting sites was governed by 2 factors: firstly, the wish to cover as broad an area as possible of the two main rivers in the area, the Mitchell and Lawley (**Fig. 1**); secondly, the need to choose sites accessible by available transport.

The Mitchell River was sampled at various sites along the main watercourse and on one of its tributaries. All of the sites were readily accessible by 4-wheel drive vehicle along well-defined tracks. The head of the Lawley River estuary was reached by boat from Walsh Point (**Fig. 1**).

In addition, trips were made by road to the King Edward River, Surveyors Pool and Crystal Creek (**Fig. 1**).

Collections were made with derris powder (7% to 8% rotenone), 10-metre seine net, dip net and handline. Visual surveys were often carried out under water with diving equipment but only one is reported (MIT-32), where no collection was made.

In most cases only a small number of specimens were retained of each species collected at a station. All specimens of unknown or rarely collected species were generally retained.

---

*Department of Ichthyology, Western Australian Museum, Francis Street, Perth, Western Australia. 6000.

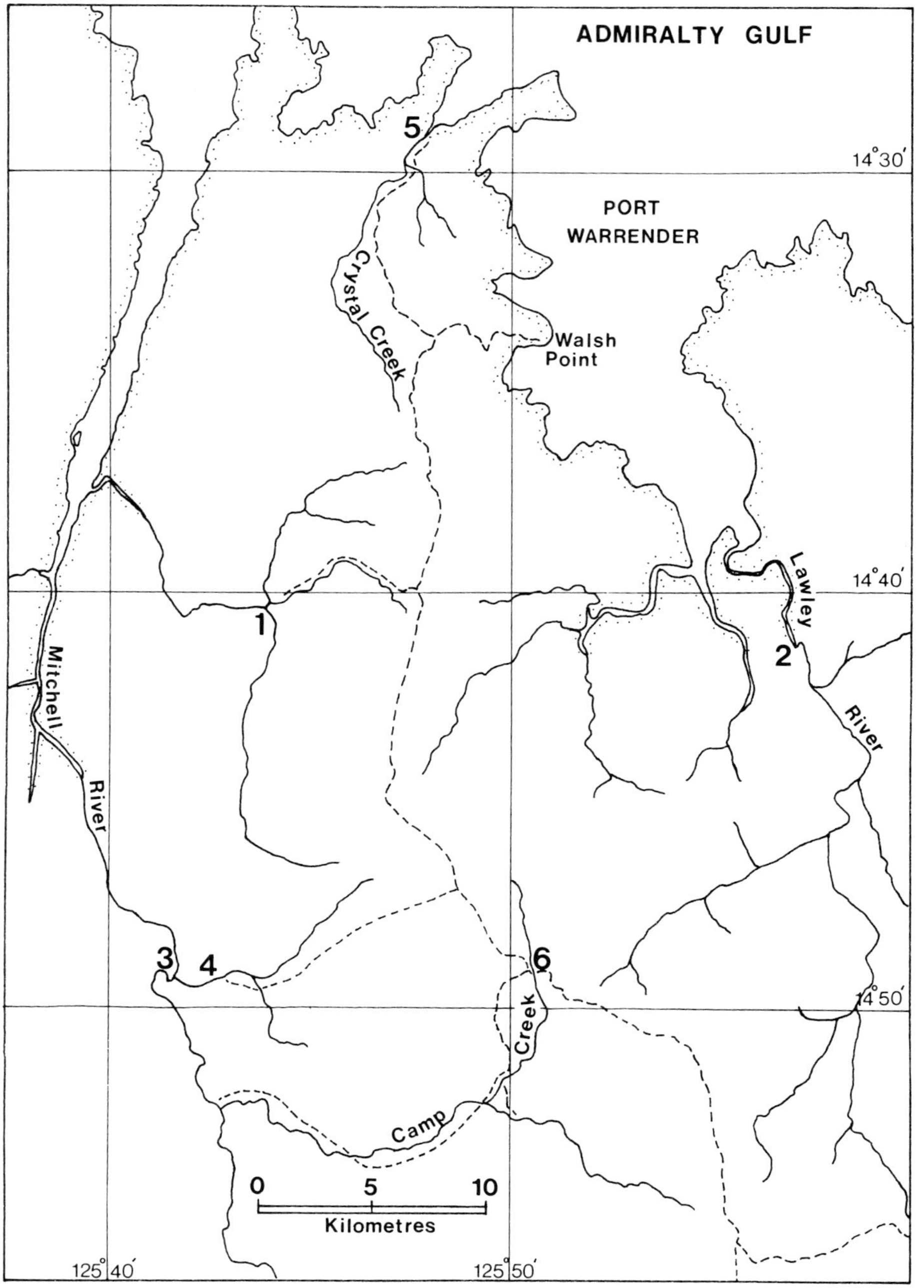

Fig. 1: Map of the Mitchell Plateau area. The following localities are indicated by numbers: **1**, Surveyors Pool; **2**, head of the Lawley River estuary; **3**, Mitchell Falls; **4**, Mertens Falls; **5**, falls on Crystal Creek; **6**, Amax camp. The King Edward River is off the map to the right.

List of Collecting Stations

All collections were made by the author, assisted by T. Duyker.

MIT-1. Camp Creek, (base camp 2), near road crossing (14°52′S, 125°50′E), 17 October 1976. Collection made in non-flowing pool (approximately 25 × 2 m) with 1 kg derris powder at depths to 2 m over sand and debris bottom; water fresh and murky.

MIT-2. Camp Creek, (base camp 2), near road crossing, 17 October 1976. Collection made in large pool (approximately 100 × 15 m) with handline at depths to 3 m over sand and debris bottom; water fresh and murky.

MIT-3. Surveyors Pool Creek, (near base camp 3), at top of falls (14°41′S, 125°44′E), 18 October 1976. Collection made in slowly flowing pool (approximately 25 × 10 m) with 1.5 kg of derris powder at depths to 1.5 m over rocky bottom; water fresh and clear.

MIT-4. Surveyors Pool, (near base camp 3), at base of falls, 18 October 1976. Collection made in large pool (approximately 100 × 60 m) with handline at depths to 3 m amongst boulders; water fresh and clear.

MIT-5. Surveyors Pool Creek, (near base camp 3), about 0.5 km downstream from falls, 19 October 1976. Collection made in slowly flowing pool (approximately 3 × 2 m) with 0.25 kg derris powder at depths to 1 m over rocky bottom; water fresh and clear.

MIT-6. Surveyors Pool Creek, (near base camp 3), about 2 km downstream from falls, 19 October 1976. Collection made in non-flowing pool (approximately 7 × 2 m) with 0.5 kg derris powder at depths to 1 m over rock and debris bottom; water fresh and clear.

MIT-17. Lawley River, (secondary camp 5), about 1 km upriver from end of mangal (14°42′S, 125°57′E), 25 October 1976. Collection made in non-flowing pool (approximately 20 × 20 m) with 1.5 kg derris powder at depths to 2 m, over rocky bottom employing stop-net across entrance to main watercourse; water fresh and clear.

MIT-18. Lawley River, (near secondary camp 5), about 0.75 km upriver from end of mangal, 25 October 1976. Collection made in non-flowing pool (approximately 30 × 3 m) with 1 kg derris powder at depths to 1.5 m over rock and debris bottom; water fresh and clear.

MIT-23. Mitchell River, about 0.5 km upriver from Mitchell Falls (14°49′S, 125°41′E), 30 October 1976. Collection made in portion (approximately 15 × 15 m) of very large non-flowing pool with 2 kg of derris powder at depths to 3 m over rock and sand bottom; water fresh and clear.

MIT-24. Mertens Creek, at base of Mertens Falls 1.5 km upstream from Mitchell Falls (14°49′S, 125°42′E), 31 October 1976. Collection made in non-flowing pool (approximately 15 × 10 m) with 1 kg derris powder at depths to 2 m over rock and sand bottom; water fresh and clear.

MIT-25. Mitchell River, about 0.25 km downriver from Camp Creek junction (14°52′S, 125°42′E), 1 November 1976. Collection made in portion of slowly flowing river (approximately 10 × 5 m) with 1 kg of derris powder at depths to 1.5 m over rock and sand bottom; water fresh and clear.

MIT-26. Mitchell River, about 0.5 km upriver from Camp Creek junction, 1 November 1976. Collection made with handline from river bank; water fresh and clear.

MIT-27. Mitchell River, about 1 km upriver from Camp Creek junction, 2 November 1976. Collection made in portion of slowly flowing river (approximately 20 × 5 m) with 1 kg derris powder at depths to 2 m over rock, weed and debris bottom; water fresh and somewhat murky.

MIT-28. King Edward River, about 2 km downriver from road crossing (14°53′S, 126°12′E), 2 November 1976. Collection made in slowly flowing pool (approximately 10 × 10 m) with 1 kg of derris powder at depths to 1 m over rocky bottom; water fresh and clear.

MIT-29. Crystal Creek, (secondary camp 1), 0.1 km downstream from falls (14°29′S, 125°48′E), 4 November 1976. Collection made in slowly flowing pool (approximately 8 × 8 m) with 1 kg derris powder at depths to 1 m over rocky bottom; water fresh and clear.

MIT-30. Crystal Creek, (near secondary camp 1), about 1 km downstream from falls, 4 November 1976. Collection made in portion (approximately 8 × 8 m) of large non-flowing pool with 2 kg derris powder in depths to 2 m over rock and debris bottom; water fresh and clear.

MIT-32. Crystal Creek, (near secondary camp 1), 0.1 km upstream from falls, 4 November 1976. Visual survey carried out underwater in slowly flowing pool (approximately 50 × 2 m) at depths to 2 m over rock and debris bottom.

RESULTS

Twenty-one species of fishes were collected from fresh water (see **Table 1**). Of these, 14 are considered to be restricted to fresh water, whereas the remaining 7 species listed below are fishes that need to return to marine or brackish waters to complete their reproductive cycles:

Megalops cyprinoides
Anguilla bicolor
Lutjanus argentimaculatus
Gerres filamentosus
Liza alata
Prionobutis microps
Hypseleotris compressus

One species is considered to be undescribed. It is listed in **Table 1** as eleotrid species.

An annotated list of species is presented below. Unless otherwise stated, all specimens are deposited at the Western Australian Museum, Perth. Catalogue numbers follow the station code numbers and measurements are standard length (SL). The type locality is enclosed in brackets following the reference to the original description for each species.

Table 1. Species content for each freshwater collecting station

									MIT								
	1	2	3	4	5	6	17	18	23	24	25	26	27	28	29	30	32
Megalops cyprinoides								X									
Anguilla bicolor																X	
Tandanus ater							X										
Tandanus glencoensis	X				X				X	X	X		X	X			
Melanotaenia australis	X		X		X	X	X	X	X	X	X		X	X	X	X	X
Ambassis agrammus	X								X				X	X			
Amniataba percoides					X	X	X	X							X		X
Hephaestus jenkinsi							X	X	X	X				X			
Leiopotherapon unicolor	X	X	X	X	X		X		X	X	X	X	X	X	X		X
Syncomistes trigonicus	X				X	X	X		X	X	X		X	X			
Glossamia aprion							X	X									
Lutjanus argentimaculatus								X								X	
Gerres filamentosus								X									
Toxotes chatareus							X	X									
Liza alata								X									
Glossogobius giurus							X	X						X			
Hypseleotris compressus								X									
Mogurnda mogurnda	X		X		X	X			X	X	X		X	X	X	X	X
Oxyeleotris lineolatus							X										
Prionobutis microps								X									
Eleotrid species									X	X							

ANNOTATED
LIST OF SPECIES

FAMILY MEGALOPIDAE — **Ox-eye Herrings**

Megalops cyprinoides

Clupea cyprinoides Broussonet, 1782: pl. 9 (Tanna Islands, New Hebrides).

MIT-18 (P25678-012), 1 specimen, 223 mm SL.

Collected from fresh water in the lower reaches of the Lawley River. Another specimen was taken but not retained. Found throughout the tropical Indo-Pacific region. In Western Australia it is known from fresh and brackish waters in the Kimberley and Pilbara regions.

FAMILY ANGUILLIDAE — **Freshwater Eels**

Anguilla bicolor

Anguilla bicolor McClelland, 1845: 178 (Bengal, India).

MIT-30 (P25690-001), 1 specimen, 493 mm SL.

This species was collected from fresh water just upstream from the mangrove area of Crystal Creek. It has also been collected from the Pilbara region of Western Australia. Known from India, Malaysia, Indonesia, Philippines, New Guinea and northern Australia.

FAMILY PLOTOSIDAE — **Eel-tail Catfishes**

Tandanus ater

Lambertia atra Perugia, 1894: 551 (Inawi, Papua).

MIT-17 (P25677-008), 4 specimens, 124-161 mm SL.

This species was collected only from the Lawley River where it is abundant. It is also found in the nearby Carson and Drysdale Rivers, as well as in the Northern Territory, Queensland and New Guinea.

Tandanus glencoensis

Copidoglanis glencoensis Rehdahl, 1922: 170 (Daly River and Glencoe, N.T.).

MIT-1 (P25661-006), 2 specimens, 76-80 mm SL.
MIT-5 (P25665-005), 2 specimens, 114-118 mm SL.
MIT-23 (P25683-005), 3 specimens, 105-121 mm SL.
MIT-24 (P25684-008), 2 specimens, 95-171 mm SL.
MIT-25 (P25685-003), 1 specimen, 81 mm SL.
MIT-27 (P25687-004), 2 specimens, 64-65 mm SL.
MIT-28 (P25688-004), 2 specimens, 52-115 mm SL.

This species was not collected from either the Lawley River or Crystal Creek but was present at all other freshwater localities visited. Its absence from the Lawley River may be explained by the presence there of *T. ater*. Distributed from the Pilbara region of Western Australia to the Northern Territory.

FAMILY MELANOTAENIIDAE **Rainbowfishes**

Melanotaenia australis

Neoatherina australis Castelnau, 1875: 32 (Western Australia [possibly Pilbara area]).

MIT-1 (P25661-003), 4 specimens, 52-62 mm SL.
MIT-3 (P25663-002), 13 specimens, 30-52 mm SL.
MIT-5 (P25665-001), 4 specimens, 36-47 mm SL.
MIT-6 (P25666-001), 7 specimens 19-41 mm SL.
MIT-17 (P25677-005), 17 specimens, 26-58 mm SL.
MIT-18 (P25678-005), 3 specimens, 24-53 mm SL.
MIT-23 (P25683-004), 2 specimens, 50-54 mm SL.
MIT-24 (P25684-005), 17 specimens, 19-66 mm SL.
MIT-25 (P25685-005), 5 specimens, 47-60 mm SL.
MIT-27 (P25687-006), 12 specimens, 46-72 mm SL.
MIT-29 (P25689-003), 23 specimens, 26-51 mm SL.
MIT-30 (P25690-003), 3 specimens, 25-35 mm SL.

Collected or observed at every freshwater collecting site. A different colour variety was found in the King Edward River. Specimens with this colour pattern (Hutchins, 1977, plate 14) from the nearby Carson and Drysdale Rivers were listed by Hutchins (1977: 104) as *Nematocentris* species 1. However, Allen (1978) has shown that *M. australis* is quite variable in colour pattern throughout its range (Pilbara, throughout the Kimberley to Darwin, N.T.).

FAMILY CENTROPOMIDAE **Glassfishes**

Ambassis agrammus

Ambassis agrammus Günther, 1867: 57 (Cape York, Qld).

MIT-1 (P25661-004), 1 specimen, 27 mm SL.
MIT-23 (P25683-007), 1 specimen, 23 mm SL.
MIT-27 (P25687-003), 9 specimens, 22-35 mm SL.
MIT-28 (P25688-008), 3 specimens, 20-25 mm SL.

Collected only from pools which contained aquatic plants. Known from freshwater streams in northern Australia and New Guinea.

FAMILY TERAPONIDAE **Grunters**

Amniataba percoides

Therapon percoides Günther, 1864: 374 (Fitzroy River, Qld).

MIT-5 (P25665-003), 1 specimen, 54 mm SL.
MIT-6 (P25666-002), 2 specimens, 24 mm SL.
MIT-17 (P25677-003), 3 specimens, 51-94 mm SL.
MIT-18 (P25678-001), 1 specimen, 79 mm SL.
MIT-29 (P25689-002), 4 specimens, 62-69 mm SL.
MIT-32, no specimens collected.

This species appears to have a scattered distribution in the Mitchell Plateau area. It was not present in either the Mitchell or King Edward Rivers at the sites sampled.

However, it was collected from Surveyors Pool Creek which runs into the long narrow estuary at the mouth of the Mitchell River (see **Fig. 1**). It was also present in Crystal Creek and the Lawley River. Widely spread throughout the rivers of northern Australia.

Hephaestus jenkinsi (**Fig. 2**)

Mesopristes jenkinsi Whitley, 1945: 26 (Ivanhoe Station, Ord River, Western Australia).

MIT-17 (P25677-001), 6 specimens, 58-132 mm SL.
MIT-18 (P25678-002), 1 specimen, 116 mm SL.
MIT-23 (P25683-001), 10 specimens, 40-159 mm SL.
MIT-24 (P25684-002), 4 specimens, 46-145 mm SL.
MIT-28 (P25688-002), 9 specimens, 35-116 mm SL.

Collected only from large pools in the Lawley, Mitchell and King Edward Rivers. Its absence from that area of the Mitchell River near Camp Creek junction was unexpected (MIT-25 to 27). Here *Leiopotherapon unicolor* is the dominant teraponid. However, the river at these sites was considerably narrower than farther downriver at MIT-17 where *H. jenkinsi* was common. One hybrid of *H. jenkinsi* and *Syncomistes trigonicus* (see below) (P25684-003, 144 mm SL) was collected from Mertens Creek (MIT-24). Another specimen of this cross (P25031-004, 145 mm SL) has been taken from the Prince Regent River. *H. jenkinsi* is common in rivers of the Kimberley region where it prefers main watercourses and large pools.

Fig. 2: *Hephaestus jenkinsi*

Leiopotherapon unicolor

Therapon unicolor Günther, 1859: 277 (Mosquito Creek, Darling Downs and Gwydir River, New South Wales).

MIT-1 (P25661-001), 5 specimens, 50-100 mm SL.
MIT-2 (P25662-001), 11 specimens, 62-100 mm SL.
MIT-3 (P25663-001), 8 specimens, 49-133 mm SL.
MIT-4 (P25664-001), 5 specimens, 85-131 mm SL.

MIT-5 (P25665-002), 1 specimen, 96 mm SL.
MIT-17 (P25677-002), 3 specimens, 68-107 mm SL.
MIT-23 (P25683-003), 1 specimen, 101 mm SL.
MIT-24 (P25684-001), 2 specimens, 71-112 mm SL.
MIT-25 (P25685-001), 1 specimen, 35 mm SL.
MIT-26 (P25686-001), 10 specimens, 150-240 mm SL.
MIT-27 (P25687-001), 3 specimens, 31-45 mm SL.
MIT-28 (P25688-001), 1 specimen, 47 mm SL.
MIT-29 (P25689-001), 5 specimens, 63-112 mm SL.
MIT-32 No specimens were collected.

Common at most of the sites visited, especially where *Hephaestus jenkinsi* was absent. Widely distributed throughout rivers of northern Australia.

Syncomistes trigonicus

Syncomistes trigonicus Vari, 1978: 316 (Wyulda Creek, Roe River, Western Australia).

MIT-1 (P25661-002), 1 specimen, 55 mm SL.
MIT-5 (P25665-004), 10 specimens, 37-77 mm SL.
MIT-6 (P25666-006), 1 specimen, 40 mm SL.
MIT-17 (P25677-004), 1 specimen, 64 mm SL.
MIT-23 (P25683-002), 11 specimens, 34-100 mm SL.
MIT-24 (P25684-004), 4 specimens, 94-127 mm SL.
MIT-25 (P25685-002), 5 specimens, 33-72 mm SL.
MIT-27 (P25687-002), 1 specimen, 50 mm SL.
MIT-28 (P25688-003), 9 specimens, 41-119 mm SL.

This species was previously reported by Allen (1975:94) as Theraponid species B and as Terapon species 2 by Hutchins (1977:106). The genus is characterised by a pointed snout and flattened teeth which are movably imbedded in the upper and lower lips, the lower series tending to project laterally. In contrast to other freshwater teraponid genera which are represented in the Kimberley region, this genus consists of species which are purely algal grazers. This species is now known to occur in the Prince Regent River, Drysdale River and Mitchell Plateau areas.

FAMILY APOGONIDAE **Cardinal-fishes**

Glossamia aprion

Apogon aprion Richardson, 1842: 16 (King's River, near Victoria, Port Essington (Darwin), Northern Territory).

MIT-17 (P25677-005), 6 specimens, 25-121 mm SL.
MIT-18 (P25678-007), 2 specimens, 48-61 mm SL.

Collected only from the Lawley River where it appears to be present in large numbers. At MIT-17 many large specimens of approximately 100 mm SL were collected but not retained. This species has a variety of colour patterns ranging from silver to purple, usually with green to brown mottlings. Its range extends from the coastal rivers of northern New South Wales to the Kimberley region.

FAMILY LUTJANIDAE **Sea Perches**

Lutjanus argentimaculatus **(Fig. 3)**

Sciaena argentimaculata Forsskål, 1775: 47 (Red Sea).

MIT-18 (P25678-003), 6 specimens, 69-97 mm SL.
MIT-30 (P25690-002), 1 specimen, 103 mm SL.

Collected from fresh water in the lower reaches of the Lawley River and Crystal Creek. It was also taken from brackish water in both these areas. In freshwater areas where this euryhaline species occurs, it dominates the freshwater species such as *Hephaestus jenkinsi* in the competition for food. However, as reported by McCristal (1976: 32), it is itself dominated by the larger *Lates calcarifer* when this euryhaline species is also present. *Lutjanus argentimaculatus* occurs in marine, brackish and freshwater environments of the Indo-West Pacific area. In Western Australia it is known from the Pilbara and Kimberley regions.

Fig. 3: *Lutjanus argentimaculatus*

FAMILY GERRIDAE **Silver-biddies**

Gerres filamentosus **(Fig. 4)**

Gerres filamentosus Cuvier, 1829: 188 (Vizagapatam [Vishakhapatnam], India).
MIT-18 (P25678-009), 2 specimens, 103-121 mm SL.

Collected only from fresh water in the lower reaches of the Lawley River. Several other specimens of a similar size were also collected but not retained. This marine species is known to ascend rivers into fresh water. It is found throughout the tropical Indo-west Pacific region, including northern Australia.

Fig. 4: *Gerres filamentosus*

FAMILY TOXOTIDAE **Archerfishes**

Toxotes chatareus

Coius chatareus Hamilton-Buchanan, 1822: 101 and 370 (Ganges River, India).
MIT-17 (P25677-010), 6 specimens, 62-106 mm SL.
MIT-18 (P25678-006), 1 specimen, 91 mm SL.

Collected and observed in both fresh water and brackish water in the lower reaches of the Lawley River. This species may be separated into 2 forms in northern Australia (Allen 1978). One inhabits the marine environment, the other prefers fresh water areas. Each form may also be found in brackish water. Apparently both forms are able to pass through their entire life cycle in their respective habitats. The distinguishing character of the freshwater form is a small spinous dorsal fin consisting of relatively slim spines, whereas the marine inhabitants possess a higher spinous dorsal containing more robust spines. This character difference may be related to differences in water composition between the marine and freshwater environments. Found in northern Australia, Malaysia, Indonesia, New Guinea and South-east Asia.

FAMILY MUGILIDAE **Mullets**

Liza alata **(Fig. 5)**

Mugil alata Steindachner, 1892: 364 (Madagascar).
MIT-18 (P25678-004), 2 specimens, 91-152 mm SL.

Collected and observed in fresh and brackish water in the lower reaches of the Lawley River. In the latter area it was observed swimming in schools of 15-20 individuals. *L. alata* was previously reported from northern Australia as *L. diadema* (Gilchrist & Thompson) by Lake (1971: 42). In Western Australia it is also known from the Ord River. The range of this species extends across to the east coast of Africa.

Fig. 5: *Liza alata*

FAMILY GOBIIDAE **Gobies**

Glossogobius giurus

Gobius giurus Hamilton-Buchanan, 1822: 51 & 366. (Ganges River, India).

MIT-17 (P25677-007), 4 specimens, 62-81 mm SL.
MIT-18 (P25678-010), 2 specimens, 44-90 mm SL.
MIT-28 (P25688-006), 2 specimens, 48-80 mm SL.

Collected from fresh water in the Lawley and King Edward Rivers. Widespread throughout northern Australia and the tropical Indo-West Pacific, apparently occurring only in fresh water.

FAMILY ELEOTRIDAE **Gudgeons**

Hypseleotris compressus **(Fig. 6)**

Eleotris compressus Krefft, 1864: 184 (Clarence River, N.S.W. and Port Denison, Qld).

MIT-18 (P25678-008), 12 specimens, 27-44 mm SL.

Collected from fresh and brackish waters in the lower reaches of the Lawley River. It is distributed from the north-west of Western Australia across northern Australia to New South Wales. In Western Australia it is also known from the Gascoyne River.

Fig. 6: *Hypseleotris compressus*

Mogurnda mogurnda

Eleotris mogurnda Richardson, 1844-8: 4 (Port Essington (Darwin), N.T.).

MIT-1 (P25661-005), 10 specimens, 23-44 mm SL.
MIT-3 (P25663-003), 11 specimens, 39-60 mm SL.
MIT-5 (P25665-006), 4 specimens, 28-36 mm SL.
MIT-6 (P25666-004), 9 specimens, 19-64 mm SL.
MIT-23 (P25683-006), 2 specimens, 40-60 mm SL.
MIT-24 (P25684-006), 2 specimens, 28-35 mm SL.
MIT-25 (P25685-004), 2 specimens, 29-44 mm SL.
MIT-27 (P25687-005), 4 specimens, 13-34 mm SL.
MIT-28 (P25688-005), 1 specimen, 17 mm SL.
MIT-29 (P25689-004), 6 specimens, 24-40 mm SL.
MIT-30 (P25690-004), 3 specimens, 19-30 mm SL.
MIT-32 No specimens collected.

Collected at all fresh water localities visited except the Lawley River where the presence of both *Glossogobius giurus* and *Oxyeleotris lineolatus*, being larger and more aggressive relatives of *M. mogurnda*, probably accounts for its absence. Its range extends from northern Australia to New Guinea.

Oxyeleotris lineolatus

Eleotris lineolatus Steindachner, 1867: 13 (Rockhampton, Queensland).

MIT-17 (P25677-009), 1 specimen, 241 mm SL.

One specimen was collected from fresh water in the Lawley River. Recorded as *Bunaka herwerdenii* (Weber), by Hutchins (1977:107) from the Carson River. Found in freshwater rivers of the northern half of Australia and New Guinea.

Prionobutis microps **(Fig. 7)**

Pogoneleotris microps Weber, 1908: 258 (Tawarin and Merauke Rivers, New Guinea).

MIT-18 (P25678-011), 2 specimens, 93-105 mm SL.

Collected from fresh and brackish waters in the lower reaches of the Lawley River, and from brackish water of Crystal Creek. Known from brackish and fresh waters of northern Australia and New Guinea.

Fig. 7: *Prionobutis microps*

Eleotrid species **(Fig. 8)**

MIT-23 (P25683-008), 2 specimens, 22-24 mm SL.
MIT-24 (P25684-007), 16 specimens, 26-32 mm SL.

Collected only from the Mitchell River in the vicinity of Mitchell Falls. This species is known only from the above records (a further 7 specimens from MIT-24 have been deposited at the Australian Museum). It is apparently undescribed and is closely related to a species reported from the Drysdale River by Hutchins (1977:107) as Gobiid species. It differs from the latter species by not possessing body bars (compare Fig. 8 with Fig. 11 in Hutchins, 1977:107). Both of these undescribed species appear to represent a new genus which has similarities with both *Hypseleotris* and *Philypnodon* but is characterized by the presence of vomerine teeth.

Fig. 8: *Eleotrid* species

DISCUSSION

The freshwater fish fauna of the Kimberley region of Western Australia was, until recently, poorly known. Papers by Allen (1975) and Hutchins (1977) have included faunal lists for the Prince Regent River area and Drysdale River area respectively. The following have also made substantial collections in the Kimberley region but these remain largely unreported: Rosen, Nelson and Butler for the American Museum of Natural History in 1969 (Gibb River Road to Drysdale River Crossing; Great Northern Highway to Wyndham); McKay for the Western Australian Museum in 1971 (Ord River area); Hutchins and Chapman for the Western Australian Museum in 1975 (Gibb River Road and Kalumburu Road to just north of Doongan Station); Allen for the Western Australian Museum in 1977 (Gibb River Road, Karungie Road to Wyndham, thence to Kununurra, Halls Creek and Fitzroy Crossing). Together with the species reported here from the Mitchell Plateau area, a rather more complete picture of the freshwater fishes of the Kimberley Region can now be drawn up. **Table 2** represents a list of the main river systems which have now been sampled in the Kimberley region, with the species known to occur in each system listed (specimens of all species are housed at the Western Australian Museum). However, much of the material collected by Rosen, Nelson and Butler is at the American Museum of Natural History and was not used for this table.

The 14 species of freshwater fishes known from the Mitchell Plateau area is comparable in number with the 15 species that were collected by Allen (1975) from the nearby Prince Regent River area, even though a helicopter was employed to get to the more inaccessible regions during the latter survey. However Hutchins (1977), also with helicopter support, collected 25 species from the Drysdale River area. While the first two areas have drainage basins of similar size, the Drysdale River area has a much greater drainage basin. This may account for the more diverse freshwater fish fauna in the latter area.

Table 2: The known freshwater fish species of the Kimberley region of Western Australia

Species	Fitzroy R.	Meda R.	Isdell R.	Prince Regent R.	Roe R.	Mitchell R.	Lawley R.	King Edward R.	Carson R.	Drysdale R.	King George R.	Berkeley R.	Ord R.
Nematalosa erebi	X	X	X	X					X	X			X
Arius australis	X							X	X				X
A. species 1[a]	X												X
A. species 2[b]	X												
A. species 3[c]			X										
Tandanus ater							X	X	X	X			
T. glencoensis	X	X	X	X	X	X		X	X	X	X	X	X
T. species 1[d]									X	X			
T. species 2[e]										X			
Belone krefftii	X								X				X
Melanotaenia australis	X	X	X	X	X	X	X	X	X	X	X	X	X
M. pygmaea				X									
M. gracilis									X	X	X	X	

a Distinguished from *A. australis* by its almost truncate snout (dorsal view) and short barbels.

b Distinguished from *A. australis* by its stocky appearance, small eye (8-9 in head length) and yellow coloration.

c Appears to be similar to *A. macronotacanthus* Bleeker, 1846 from Indonesia and Malaysia.

d Recorded as listed by Hutchins, 1977:104.

e Recorded as listed by Hutchins, 1977:104.

Table 2: (continued)

Species	Fitzroy R.	Meda R.	Isdell R.	Prince Regent R.	Roe R.	Mitchell R.	Lawley R.	King Edward R.	Carson R.	Drysdale R.	King George R.	Berkeley R.	Ord R.
Craterocephalus majoriae										X			
C. species[f]					X								
Quirichthys stramineus													X
Ambassis agrammus	X	X	X	X		X		X	X	X	X	X	X
A. macleayi									X				
Parambassis gulliveri													X
Amniataba percoides	X						X	X	X	X			X
Hannia greenwayi	X												
Hephaestus jenkinsi	X		X	X		X	X	X	X	X		X	X
H. epirrhinos									X	X			
Leiopotherapon unicolor	X	X	X	X	X	X	X	X	X	X	X	X	X
L. macrolepis				X	X								
Syncomistes trigonicus				X	X	X	X	X	X	X			
S. butleri										X			X
S. kimberleyensis													X
S. rastellus										X			

f Recorded as *C. worrelli* by Allen, 1975: 92.

Table 2: (continued)

Species	Fitzroy R.	Meda R.	Isdell R.	Prince Regent R.	Roe R.	Mitchell R.	Lawley R.	King Edward R.	Carson R.	Drysdale R.	King George R.	Berkeley R.	Ord R.
Glossamia aprion	X						X		X				X
Toxotes chatareus							X		X				X
T. oligolepis[g]	X	X	X										
Glossogobius aureus[h]				X									
G. giurus	X	X				X	X	X	X	X			X
Hypseleotris compressus							X						
H. species 1[i]	X												
H. species 2[j]					X								
H. species 3[k]				X									
Mogurnda mogurnda			X	X	X	X		X	X	X	X	X	X
M. species[l]					X								
Oxyeleotris lineolatus	X						X		X				X
Eleotrid species 1[m]										X			
Eleotrid species 2[n]						X							

g Recorded as an undescribed species of *Toxotes* by Allen, 1975: 94.

h Recorded as *G. giurus* by Allen, 1975: 94.

i, j and k Being described as new by Hoese (see Acknowledgements).

l An apparently undescribed species which differs from *M. mogurnda* in longitudinal scale count (30-36 as against 37-46).

m Recorded as Gobiid species by Hutchins, 1977: 107. It is being described as a new genus and species by Hoese.

n Discussed within this report as Eleotrid species.

ACKNOWLEDGEMENTS

The author wishes to thank the following persons for assistance with the identification of species in their respective speciality: G.R. Allen, Western Australian Museum, Perth (Melanotaeniidae and Toxotidae); D.F. Hoese, Australian Museum, Sydney (Eleotridae); J.M. Thompson, University of Queensland, Brisbane (Mugilidae); and R.P. Vari, National Museum of Natural History, Washington, D.C. (Teraponidae).

REFERENCES

ALLEN, G.R. (1975)—A preliminary checklist of the freshwater fishes of the Prince Regent River Reserve, north-west Kimberley, Western Australia. *In* Miles, J.M. & Burbidge, A.A. (Eds) A biological survey of the Prince Regent River Reserve, north-west Kimberley, Western Australia in August, 1974. *Wildl. Res. Bull. West. Aust.* no. 3: 89-96.

ALLEN, G.R. (1978)—The rainbowfishes (family Melanotaeniidae) of north-western Australia. *Trop. Fish Hobbyist* **1978** (June) 91-102.

ALLEN, G.R. (1978)—A review of the archerfishes (family Toxotidae). *Rec. West. Aust. Mus.* **6:** 355-378.

BROUSSONET, P.M.A. (1782)—*Ichthyologia, sistens piscium descriptiones et icones.* London: Elmsly.

CASTELNAU, F.L. (1875)—Researches on the fishes of Australia. Philad. Centen. Exhib. 1876 (Melbourne 1875): Official Record. *Intercolonial Exhibition Essays* no. 2: 1-52.

CUVIER, G. (1829)—*Le règne animal* . . . . 2nd.ed. vol.2. Paris.

FORSSKÅL, P. (1775)—*Descriptiones animalium* . . . . Havniae.

GÜNTHER, A. (1859)—*Catalogue of the acanthopterygian fishes in the collection of the British Museum.* **1.** London: The Trustees of the British Museum (Natural History).

GÜNTHER, A. (1864)—Descriptions of three new species of fishes in the collection of the British Museum. *Ann. Mag. nat. Hist.* (3) **14:** 374-376.

GÜNTHER, A. (1867)—Additions to the knowledge of Australian reptiles and fishes. *Ann. Mag. nat. Hist.* (3) 20: 45-68.

HAMILTON-BUCHANAN, F. (1822)—*An account of the fishes found in the river Ganges and its branches.* London.

HUTCHINS, J.B. (1977)—The freshwater fish fauna of the Drysdale River National Park, north Kimberley, Western Australia. *Wildl. Res. Bull. West. Aust.* no. 6: 102-109.

KREFFT, G. (1864)—Notes on Australian freshwater fishes and descriptions of four new species. *Proc. zool. Soc. Lond.* **1864:** 182-184.

LAKE, J.S. (1971)—Freshwater fishes and rivers of Australia. Melbourne: Thomas Nelson.

McCLELLAND, J. (1845)—Apodal fishes of Bengal. *Calcutta J. nat. Hist.* **5:** 150-226.

McCRISTAL, V. (1976)—Freshwater fighting fish. *Aust. Outdoors Fishing,* **1976**(November): 28-32; 66-68.

MEES, G.F. (1962)—Additions to the fish fauna of Western Australia pt. 3. *Fish. Bull. West. Aust.* **9**(3): 23-30.

PERUGIA, A. (1894)—Viaggio di Lamberto Loria nella Papuasia orientale. Pesci d'acqua dolce. *Ann. Mus. Civ. Stor. nat. Giacomo Doria.* (2) **14:** 546-553.

RENDAHL, H. (1922)—A contribution to the ichthyology of north-west Australia. *Meddr. zool. Mus., Oslo* **5:** 163-197.

RICHARDSON, J. (1842)—Contributions to the ichthyology of Australia. *Ann. Mag. nat. Hist.* **9:** 15-31.

RICHARDSON, J. (1844-8)—*Ichthyology of the voyage of H.M.S. Erebus and Terror.* London: Newman.

STEINDACHNER, F. (1867)—Ueber einige Fische aus dem Fitzroy-Flusse bei Rockhampton in Ost-Australien. *Sitzber. Akad. Wiss. Wien.* **55:** 9-16.

STEINDACHNER, F. (1892)—Ueber einige neue and seltene Fischarten aus der ichthyologischen Sammlung des K. K. Naturhistorischen Hofmuseums. *Denkschr. Akad. Wiss. Wien* **59:** 357-384.

VARI, R.P. (1978)—The Terapon perches (Percoidei; Teraponidae), a cladistic analysis and taxonomic revision. *Bull. Amer. Mus. nat. Hist.* **159:** 175-340.

WEBER, M. (1908)—Süsswasserfische von Neu-Guinea . . . . Nova Guinea **5:** 201-267.

WHITLEY, G.P. (1945)—New sharks and fishes from Western Australia. Pt.2. *Aust. Zool.* **11:** 1-42.

Received 28 February 1978 Accepted 30 March 1978 Published 28 February 1981

PART 8

MOLLUSCAN FAUNA OF THE ADMIRALTY GULF, CAPE VOLTAIRE, AND THE INSTITUT ISLANDS, KIMBERLEY, WESTERN AUSTRALIA. CHITONS, MESO- AND NEOGASTROPODS.

F. E. WELLS*

INTRODUCTION

The Admiralty Gulf **(Fig. 1)** is a large marine embayment with an area of approximately 1700 km^2 on the extreme north-east coast of Western Australia, about 300 km west of the Northern Territory border. The waters of the Gulf are essentially marine but small amounts of freshwater flow in from the Mitchell and Lawley Rivers; limited estuarine areas are associated with each river. The complex system of tides in the Gulf makes the area an interesting one for examining patterns of zonation of marine intertidal mollusc species. The tidal range is extensive, with a maximum range of as much as 8.8 m on spring tides, from -1.7 m on the shore to +7.1 m. The average tidal amplitude is 4.4 m. The tides are predominately semidiurnal, but diurnal tides occur during the changeover between neap and spring tide periods. When the changeover occurs the tidal levels remain static for several hours at a level of about +2.7 m.

The Gulf is bounded on the east by Cape Bougainville and on the west by Cape Voltaire. The Institut Islands are a small archipelago north of Cape Voltaire. The water of the islands is from the open sea and lacks the heavy silt content of the water of the Admiralty Gulf. The islands are mostly small, less than 5 km long, and only the largest were visited on this study. Well developed coral reefs occur in the shallows offshore from the islands but are less well developed within the Admiralty Gulf itself.

METHODS

Two field programmes were undertaken in the Admiralty Gulf during 1976. The first by this author was a brief survey of the Admiralty Gulf, Cape Voltaire and the Institut Islands made from 30 July to 20 August. The first 10 days of this trip were devoted to examining fauna of the area around Port Warrender in the Admiralty Gulf and planning the detailed mangrove survey to be conducted on the second trip. During this initial period Middle Osborn Island was investigated. After completing

*Western Australian Museum, Francis Street, Perth, Western Australia. 6000.

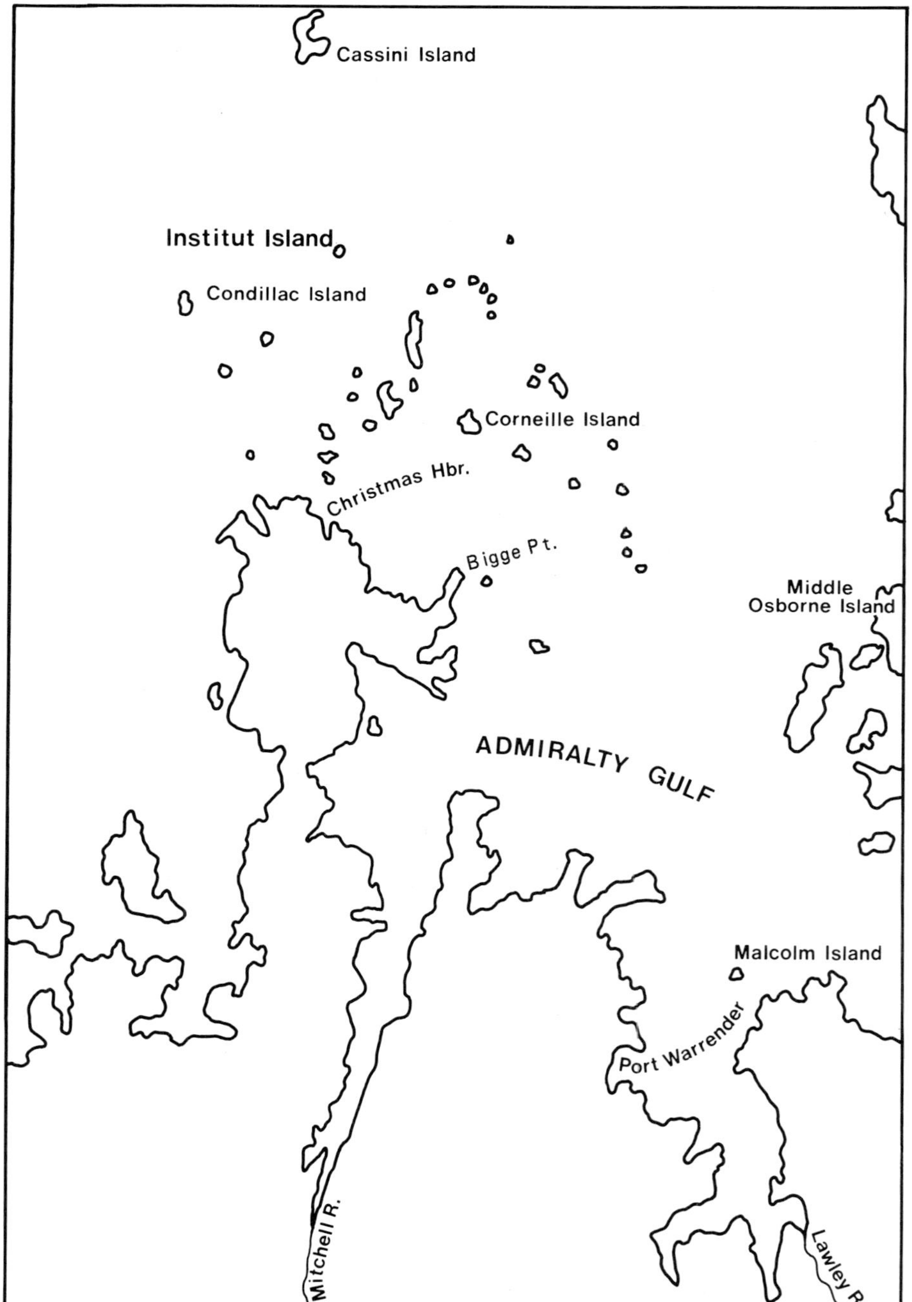

Fig. 1: Map of the Admiralty Gulf and Institut Islands showing the locations of the study sites.

the preliminary survey around Port Warrender, a 10-day survey of Cape Voltaire and the Institut Islands was made in conjunction with personnel from the Aboriginal Sites and Maritime Archaeology departments of the Western Australian Museum. Two 5 m boats equipped with outboard motors were used. The survey provided time for a brief examination of various habitats such as rock platforms and coral reefs which do not occur in the Gulf itself.

The second expedition, which lasted from 10 October to 6 November, was a detailed examination of the ecology of marine molluscs. It was undertaken at Port Warrender by this author with S.M. Slack-Smith and C.W. Bryce, also of the Western Australian Museum. Two other studies were undertaken at Port Warrender: an examination of the zonation of gastropods on the rocky shore at the tip of Walsh Point (Wells, 1978) and a detailed study of the molluscs in the mangal at Port Warrender (Wells & Slack-Smith, Part 9, this publication). These studies also provided specimens used in the faunal survey. No collecting was undertaken outside the Port Warrender area on the October trip.

Two basic types of collecting were undertaken on each trip. During low tides the shoreline was examined for living specimens. At high tide the beach drift was searched for species not collected alive. The water around Port Warrender is too shallow and muddy for diving, but snorkel and Scuba were used at Middle Osborn Island, Christmas Harbour and Cassini Island. No dredging was done at any station. Specimens collected alive were preserved initially in 10% formalin and were transferred to 70% alcohol after the expedition. The specimens were identified at the Western Australian Museum and representative specimens of all species collected have been retained.

Collections were made at several sites in the Admiralty Gulf, Cape Voltaire and the Institut Islands as follows:

Port Warrender (PW). The area around Port Warrender is typical of the Admiralty Gulf and the species list from this station appears to be representative of other areas within the Gulf. Two main habitats occur at Port Warrender: rocky shores on the headlands, interspersed with deeply indented bays lined with mangrove swamps and extensive mudflats.

Walsh Point is a basaltic peninsula 2.7 km long. The seaward tip, where most collecting was done, is almost completely rocky in the intertidal areas. The high intertidal zone has rocks 5-13 cm in diameter and several layers deep. These rocks are easily shifted about by even moderate wave action. The rocks are 10-30 cm in diameter at the mid-tide level and are piled only 2 or 3 deep. Baserock is exposed in some areas of the mid-tide zone. At the lower intertidal areas the rocks are 30 cm or more in diameter and separated by as much as 30 cm of sand. The tip of Walsh Point changes abruptly into a sandy beach at about the 0.5 m level. Along the north and south sides of Walsh Point the lower beach is composed of mud rather than sand.

The mangal studied in detail at Port Warrender is located at the head of Port Warrender Bay, a small bay to the south of Walsh Point. The mangal extends south in a band about 300 m wide for 3.2 km to the next unnamed headland at a mid-tide level of +2.9 to +5.2 m. Details of the zonation of mangroves and molluscs are discussed elsewhere (Wells & Slack-Smith, Part 9, this publication). A tidal flat extends 0.6 km seaward in front of the mangal at low tide, and an extensive mudflat behind it is 0.8 km wide. The only major tidal channel through the mangal is at the southern end. There is no fresh water flowing into the Port Warrender Bay mangal except during and immediately after rainstorms. Some estuarine species collected at the mouth of the Lawley River 10 km from Port Warrender Bay are included in the list from Port Warrender.

Port Warrender is located at the head of the Admiralty Gulf and thus is well protected from the wave action of open sea. The water is extremely turbid at low tide even on calm days, with a visibility of 15 cm or less. As the tide rises the visibility increases to about 1 m.

Malcolm Island (MI) is a small island in the Admiralty Gulf north-east of the Port Warrender area. It was visited on 31 October, when mid-tide collecting was carried out on intertidal sandstone rocks and from drift on the sandy beach.

Middle Osborn Island (MOI), Admiralty Gulf, was visited on day trips on 1 and 6 August. During low tide, collecting was done on a shoreline composed of a mixture of mangroves and rock boulders. As the tide rose, collecting shifted to the beach drift higher up the shore.

Bigge Point (BP), east of Cape Voltaire, was examined briefly on a rest stop during the afternoon of the 11 August trip from Port Warrender to Christmas Harbour. At the head of the bay was a coarse sand beach. The sides of the bay were lined with sandstone rock.

Christmas Harbour (CH), Cape Voltaire, served as the base camp from 11 to 13 August. The campsite was located on a sand beach which had fine grained sand in the low tide area. The northern side of the beach was composed of a solid sandstone outcrop lined to seaward with boulders that had broken away. The southern side of the beach had boulders as much as 1 m in diameter but these did not reach the low tide level. During low tides of 0.0 m a small fringing coral reef about 20 m offshore became exposed and collections were made there. Water in the area had a high sediment content and corals were not well developed. Additional collections were made below the low tide level, to a depth of 6 m. Beach drift specimens were obtained on the exposed side of Cape Voltaire during high tides.

Condillac Island (CI), Institut Islands, was studied on the night of 14-15 August. Collections were made on a basaltic rock platform during two excellent low tides. This station provided the only substantial rock-platform collecting during either trip. Additional specimens were obtained from beach drift.

Corneille Island (CNI), Institut Islands, was visited from 16 to 18 August. The area investigated was a sandy beach at the south-eastern corner of the island. The bay was lined on both sides with basaltic bedrock and there were few boulders in the intertidal zone which could be turned over. A small collection of beach drift specimens from Lagrange Island 2 km away is included in the list for Corneille Island.

Cassini Island (CSI), Institut Islands, was studied on 19 August. During the morning, collections were made on a well developed coral reef off the southern end of the island. The island is basaltic with a sandy beach on the east coast. This area was visited during the afternoon and collections were made along the rocky shore and from the beach drift.

ANNOTATED LIST OF SPECIES

The following pages list the species collected during the study. The list is arranged in taxonomic order by classes, orders and families. Genera and species are arranged alphabetically within families. The localities where a species was collected are listed in the order in which they were visited. Lists include localities where live-collected specimens were obtained as well as those where only beach drift material was found. In a study of this type it is impossible, and probably unnecessary, to provide a taxonomic basis for the identification of each species; many of the identifications are tentative. Instead, wherever possible reference is made to a published figure of the species found in the Admiralty Gulf. Supplementary notes are provided where they supply pertinent additional information.

PHYLUM MOLLUSCA

CLASS POLYPLACOPHORA

FAMILY CHITONIDAE

Acanthopleura gemmata Blainville, 1825. Localities: PW, MOI, CH, CNI, CI, CSI. Figured by Iredale and Hull (1927). Plate XV, Fig. 33.

Acanthopleura spinosa Bruguière, 1792. Localities: PW, MOI, CH, CNI, CI, CSI. Figured by Iredale and Hull (1927). Plate XVI, Fig. 1.

CLASS GASTROPODA

SUBCLASS PROSOBRANCHIA

ORDER ARCHAEOGASTROPODA

FAMILY NERITIDAE

Nerita albicilla Linnaeus, 1758. Localities: PW, BP, CNI, CI, CH. Figured by Wilson and Gillett (1974), 30: Plate 11, no. 3.

Nerita chamaeleon Linnaeus, 1758. Localities: PW, CH. Figured by Wilson and Gillett (1974), 30: Plate 11, no. 8.

Nerita lineata Gmelin, 1791. Localities: PW. Figured by Wilson and Gillett (1974), 30: Plate 11, no. 7.

Nerita polita Linnaeus, 1758. Localities: PW, CH. Figured by Wilson and Gillett (1974), 30: Plate 11, no. 1.

Nerita reticulata Karsten, 1789. Localities: PW, BP. Figured by Cernohorsky (1972a) 265: Plate 11, Fig. 10.

Nerita undata Linnaeus, 1758. Localities: PW, BP, CH, CNI, CI. Figured by Wilson and Gillett (1974) 30: Plate 11, no. 4.

Neritina cf. *oualaniensis* Lesson, 1830. Localities: PW. Figured by Coleman (1975) 32: Fig. 60.

Neritina sp. Localities PW. Remarks: This species was collected in a small tidepool in the mangrove swamp at Port Warrender.

ORDER MESOGASTROPODA

FAMILY LITTORINIDAE

Littorina scabra scabra (Linnaeus, 1758). Localities: PW. Figured by Rosewater (1970) page 918: Plate 325, Figs 1-4, page 457: Plate 352, Figs 1-34.

Nodilittorina pyramidalis pyramidalis (Quoy and Gaimard, 1833). Localities: CI. Figured by Rosewater (1970) 418: Plate 325, Figs 24-25; page 420: Plate 326, Figs 9 and 11; page 481: Plate 370, Figs 1-13.

Tectarius pagodus (Linnaeus, 1758). Localities: CH, CNI, CSI. Figured by Rosewater (1972) page 508: Plate 388, Figs 5-7.

FAMILY TURRITELLIDAE

Turritella terebra (Linnaeus, 1758). Localities: PW. Figured by Coleman (1975) page 244: Fig. 675.

FAMILY VERMETIDAE

Serpulorbis sp. Localities: PW, CSI.

FAMILY SILIQUARIDAE

Siliquaria ponderosa Mörch, 1860. Localities: PW, CNI. Figured by Rippingale and McMichael (1961) 46: Plate 4, no. 17.

FAMILY PLANAXIDAE

Planaxis sulcatus (Born, 1780). Localities: PW, BP, CH, CNI, CI. Figured by Cernohorsky (1972a) 267: Plate 12, Fig. 17.

FAMILY MODULIDAE

Modulus tectum (Gmelin, 1791). Localities: CI. Figured by Cernohorsky (1972a) page 267: Plate 12, Fig. 22.

FAMILY POTAMIDIDAE

Cerithidea obtusa (Lamarck, 1822). Localities: PW, MOI, CNI. Figured by Cernohorsky (1972a), 269: Plate 13, Fig. 1.

Telescopium telescopium (Linnaeus, 1758). Localities: PW, MOI. Figured by Wilson and Gillett (1974), 32: Plate 12, no. 1.

Terebralia sulcata (Born, 1778). Localities: PW. Figured by Wilson and Gillett (1974), 32: Plate 12, Fig. 2.

FAMILY CERITHIIDAE

Bittium zebrum (Kiener, 1841). Localities: CH. Figured by Cernohorsky (1972a), 271: Plate 14, Fig. 10.

Cerithium aluco (Linnaeus, 1758). Localities: CH. Figured by Cernohorsky (1972a), 269: Plate 13, Fig. 9.

Cerithium cumingi A. Adams, 1855. Localities: CI. Figured by Wilson and Gillett (1974), 32: Plate 12, Fig. 2.

Cerithium echinatum (Lamarck, 1822). Localities: PW, CNI, CI, CSI. Figured by Cernohorsky (1972a), 269: Plate 13, Fig. 12.

Cerithium novaehollandiae Sowerby, 1855. Localities: CH, CNI. Figured by Cernohorsky (1972a), 269: Plate 13, Fig. 11.

Clypeomorus batillariaeformis Habe and Kosuge, 1966. Localities: PW, MOI, CH, CNI. Figured by Cernohorsky (1972a), 275: Plate 16, Fig. 11.

Clypeomorus traillii (Sowerby, 1855). Localities: PW, CH, CNI. Figured by Cernohorsky (1972a), 275: Plate 16, Fig. 10.

Clypeomorus zonatus (Wood, 1828). Localities: PW, CH, CNI. Figured by Cernohorsky (1972a), 275: Plate 16, Fig. 9.

cf. *Colina* sp. Localities: PW.

Rhinoclavis brettinghami Cernohorsky, 1974. Localities: PW, CH, CNI, CSI. Figured by Cernohorsky (1974) 122: Figs 1, 2.

FAMILY CALYPTRAEIDAE

Siphopatella walshi Reeve. Localities: PW.

FAMILY HIPPONICIDAE

Cheilea equestris (Linnaeus, 1758). Localities: CH. Figured by Cernohorsky (1972a), 287: Plate 22, Fig. 2.

FAMILY STROMBIDAE

Lambis lambis Linnaeus, 1758. Localities: CNI, CI, CSI. Figured by Abbott (1960b), 149: Plate 121, Fig. 4; 147: Plate 118; 151: Plate 123; Figs 1-3. Figured by Wilson and Gillett (1974), 36: Plate 14, Fig. 1.

Strombus urceus Linnaeus, 1758. Localities: PW, MOI, CH, CNI, CSI. Figured by Abbott (1960a) 42: Plate 20, Figs 26-29; 63; Plate 41, Figs 1-6.

Strombus vittatus campbelli Griffith and Pidgeon, 1834. Localities: PW, MOI, CH, CNI. Figured by Abbott (1960a), 38: Plate 17, Fig. 13.

FAMILY CYPRAEIDAE

Cypraea annulus Linnaeus, 1758. Localities: CH, CI, CSI. Figured by Wilson and Gillett (1974), 56: Plate 35, no. 5.

Cypraea arabica Linnaeus, 1758. Localities: CH. Figured by Wilson and Gillett (1974), 50: Plate 30, no. 5.

Cypraea argus Linnaeus, 1758. Localities: CH, CI, CSI. Figured by Wilson and Gillett (1974), 48: Plate 28, no. 4.

Cypraea caputserpentis Linnaeus, 1758. Localities: CH. Figured by Wilson and Gillett (1974), 54: Plate 34, no. 5.

Cypraea carneola Linnaeus, 1758. Localities: CH. Figured by Wilson and Gillett (1974), 50: Plate 30, no. 9.

Cypraea clandestina Linnaeus, 1767. Localities: CI. Figured by Wilson and Gillett (1974), 56: Plate 35, no. 15.

Cypraea cylindrica Born, 1778. Localities: PW, CH. Figured by Wilson and Gillett (1974), 52: Plate 32, no. 7.

Cypraea eglantina Duclos, 1833. Localities: CI, CSI. Figured by Wilson and Gillett (1974), 50: Plate 29, no. 1.

Cypraea erosa Linnaeus, 1758. Localities: CH. Figured by Wilson and Gillett (1974), 52: Plate 32, no. 3.

Cypraea errones Linnaeus, 1758. Localities: PW, MOI, CH, CNI. Figured by Wilson and Gillett (1974), 52: Plate 32, no. 5.

Cypraea gracilis Gaskoin, 1849. Localities: PW, CH, CSI. Figured by Wilson and Gillett (1974), 56: Plate 35, no. 12.

Cypraea lynx Linnaeus, 1758. Localities: CH, CNI. Figured by Wilson and Gillett (1974), 50: Plate 30, no. 8.

Cypraea moneta Linnaeus, 1758. Localities: PW, CNI. Figured by Wilson and Gillett (1974), 56: Plate 35, no. 2.

Cypraea nucleus Linnaeus, 1758. Localities: CSI. Figured by Wilson and Gillett (1974), 58: Plate 38, no. 3.

Cypraea pyriformis Gray, 1824. Localities: PW. Figured by Wilson and Gillett (1974), 54: Plate 34, no. 11.

Cypraea quadrimaculata Gray, 1824. Localities: CH, CSI. Figured by Wilson and Gillett (1974), 56: Plate 35, no. 7.

Cypraea talpa Linnaeus, 1758. Localities: CI, CSI. Figured by Wilson and Gillett (1974), 48: Plate 28, no. 5.

Cypraea vitellus Linnaeus, 1758. Localities: CH, CNI. Figured by Coleman (1975), 129: Fig. 391.

FAMILY AMPHIPERATIDAE

Aclyvolva haynesi Sowerby, 1889. Localities: PW.

Volva philippinarium (Sowerby, 1849). Localities: PW. Figured by Cernohorsky (1972a), 288: Plate 23, Fig. 3.

FAMILY NATICIDAE

Mammilla opaca (Recluz, 1851). Localities: CNI, CSI. Figured by Wilson and Gillett (1974), 65: Plate 45, no. 6.

Mammilla simiae (Deshayes, 1838). Localities: CSI. Figured by Wilson and Gillett (1974), 64: Plate 45, no. 8.

Natica euzona Recluz, 1844. Localities: PW, CH, CSI. Figured by Cernohorsky (1972a), 293: Plate 25, Fig. 1.

Natica gualtieriana Recluz, 1844. Localities: PW, CNI. Figured by Cernohorsky (1972a), 291: Plate 24, Fig. 9.

Natica onca (Röding, 1798). Localities: CH. Figured by Cernohorsky (1972a), 293: Plate 25, Fig. 6.

Natica solida (Blainville, 1825). Localities: MOI, CH, CNI. Figured by Wilson and Gillett (1974), 64: Plate 45, no. 10.

Notocochlis sagittata Menke, 1843. Localities: CSI. Figured by Wilson and Gillett (1974), 65: Plate 45, no. 12.

Polinices pyriformis Recluz, 1844. Localities: CSI. Figured by Wilson and Gillett (1974), 64: Plate 45, no. 3.

FAMILY CASSIDAE

Casmaria erinaceus erinaceus Linnaeus, 1758. Localities: CSI. Figured by Abbott (1968), 30: Plate 14, Figs 7-12.

FAMILY FICIDAE

Ficus tessellata Kobelt, 1881. Localities: CSI. Figured by Wilson and Gillett (1974), 72: Plate 49, no. 8.

FAMILY CYMATIIDAE

Gutturnium muricinum (Röding, 1798). Localities: CSI. Figured by Coleman (1975), 53: Fig. 134.

Gyrineum pusillum (Broderip, 1832). Localities: PW. Figured by Cernohorsky (1972a), 307: Plate 32, Fig. 4.

Ranularia pyrum (Linnaeus, 1758). Localities: CSI. Figured by Wilson and Gillett (1974), 76: Plate 52, no. 5.

Septa pileare (Linnaeus, 1758). Localities: CSI. Figured by Coleman (1975), 117: Fig. 350.

FAMILY BURSIDAE

Bursa rana (Linnaeus, 1758). Localities: PW. Figured by Wilson and Gillett (1974), 80: Plate 54, no. 7.

ORDER NEOGASTROPODA

FAMILY MURICIDAE

Bedeva blosvillei (Deshayes, 1832). Localities: PW. Figured by Radwin and D'Attilio (1976). Plate 2, Fig. 8.

Chicoreus brunneus (Link, 1807). Localities: CSI. Figured by Radwin and D'Attilio (1976). Plate 4, Fig. 11.

Chicoreus cornucervi (Röding, 1798). Localities: PW, MOI, CH. Figured by Radwin and D'Attilio (1976). Plate 4, Fig. 11.

Chicoreus microphyllus (Lamarck, 1816). Localities: CSI. Figured by Radwin and D'Attilio (1976). Plate 4, Fig. 7; Plate 5, Fig. 7.

Chicoreus penchinati (Crosse, 1861). Localities: CI. Figured by Radwin and D'Attilio (1976). Plate 6, Fig. 4.

Homalocantha secunda (Lamarck, 1822). Localities: MOI, CH. Figured by Radwin and D'Attilio (1976). Plate 8, Figs 1, 2, 5.

Murex brevispinus Lamarck, 1822. Localities: PW. Figured by Radwin and D'Attilio (1976). Plate 11, Fig. 2.

Murex copperingi E.A. Smith, 1884. Localities: PW. Figured by Radwin and D'Attilio (1976). Plate 11, Fig. 1.

Murex cervicornis Lamarck, 1822. Localities: PW. Figured by Radwin and D'Attilio (1976). Plate 11, Fig. 5.

Naquetia capucina (Lamarck, 1822). Localities: PW, MOI, BP, CH. Figured by Radwin and D'Attilio (1976). Plate 15, Fig. 13.

Pterochelus acanthopterus (Lamarck, 1816). Localities: MOI. Figured by Radwin and D'Attilio (1976). Plate 15, Figs 1-3.

Pterynotus bednalli (Brazier, 1878). Localities: CH. Figured by Radwin and D'Attilio (1976). Plate 9, Figs 4, 8.

Pterynotus triquetra (Born, 1778). Localities: CSI. Figured by Radwin and D'Attilio (1976). Plate 15, Fig. 11.

FAMILY THAIDIDAE

Cronia avellana (Reeve, 1846). Localities: PW, CH, CNI, CI. Figured by Wilson and Gillett (1974), 92: Plate 61, no. 12.

cf. *Cuma gradata* (Jonas). Localities: PW. Figured by Habe and Kosuge (1965), 55: Plate 20, no. 10.

Drupella cornus (Röding, 1798). Localities: CH. Figured by Wilson and Gillett (1974), 94: Plate 61, no. 14.

Morula margariticola (Broderip, 1832). Localities: PW, MOI, CH, CNI, CI, CSI. Figured by Wilson and Gillett (1974), 92: Plate 61, no. 9.

Morula marginatra Blainville. Localities: CH. Figured by Habe and Kosuge (1965), 55: Plate 20, no. 9.

Morula nodicostata (Pease, 1868). Localities: CH, CNI. Figured by Cernohorsky (1972a), 315: Plate 36, Fig. 5.

Morula spinosa (H. and A. Adams, 1853). Localities: PW, CI, CSI. Figured by Wilson and Gillett (1974), 92: Plate 61, no. 6.

Thais aculeata Deshayes, 1844. Localities: CH. Figured by Wilson and Gillett (1974), 90: Plate 60, no. 5.

Thais echinata (Blainville, 1832). Localities: PW, MOI, CH, CI, CSI. Figured by Wilson and Gillett (1974), 90: Plate 60, no. 7.

Thais kieneri Deshayes, 1844. Localities: PW, MOI, BP, CNI. Figured by Wilson and Gillett (1974), 90: Plate 60, no. 6.

FAMILY COLUMBELLIDAE

Columbella duclosiana. Localities: PW.

Pyrene essingtonensis (Reeve, 1858). Localities: BP, CH. Figured by Habe and Kosuge (1965), 60: Plate 21, nos 1, 2.

Pyrene scripta (Lamarck, 1822). Localities: PW, CH. Figured by Cernohorsky (1972a), 323: Plate 40, Fig. 8.

Pyrene testudinaria (Link, 1807). Localities: PW, CI, CSI. Figured by Cernohorsky (1972a), 323: Plate 40, Fig. 4.

FAMILY BUCCINIDAE

Cantharus erythrostoma Reeve, 1846. Localities: PW, CH, CNI. Figured by Wilson and Gillett (1974), 96: Plate 63, no. 4.

Colubraria sp. Localities: CH.

Cominella acutinodosa Reeve, 1846. Localities: PW.

FAMILY MELONGENIDAE

Syrinx aruanus (Linnaeus, 1758). Localities: CH, CNI. Figured by Wilson and Gillett (1974), 98: Plate 63, no. 8.

Volegalea wardiana Iredale, 1938. Localities: PW, MOI. Figured by Wilson and Gillett (1974), 98: Plate 63, no. 7.

FAMILY NASSARIIDAE

Nassarius algida. Localities: CNI.

Nassarius bicallosa (E.A. Smith, 1876). Localities: CNI. Figured by Cernohorsky (1972b), 169: Figs 119, 120.

Nassarius concinnus (Powys, 1835). Localities: PW. Figured by Cernohorsky (1972b), 176: Figs 132, 133.

Nassarius dorsatus (Röding, 1798). Localities: PW, MOI. Figured by Wilson and Gillett (1974), 102: Plate 66, no. 8.

Nassarius glans (Linnaeus, 1758). Localities: PW, CNI. Figured by Wilson and Gillett (1974), 102: Plate 66, no. 11.

Nassarius olivaceus (Bruguière, 1789). Localities: BP. Figured by Cernohorsky (1972a), 329: Plate 43, Fig. 7.

FAMILY FASCIOLARIIDAE

Latirus belcheri (Reeve, 1847). Localities: CH, CI, CSI. Figured by Wilson and Gillett (1974), 100: Plate 65, no. 1.

Latirus paetelianus Kobelt, 1876. Localities: PW, MOI. Figured by Wilson and Gillett (1974), 100: Plate 65, no. 6.

Peristernia incarnata (Kiener, 1840). Localities: CNI, CI, CSI. Figured by Wilson and Gillett (1974), 100: Plate 65, no. 10.

Pleuroploca filamentosa (Röding, 1798). Localities: CI. Figured by Wilson and Gillett (1974), 98: Plate 64, no. 3.

FAMILY OLIVIDAE

Ancillista cingulata (Sowerby, 1830). Localities: CNI. Figured by Wilson and Gillett (1974), 108: Plate 71, no. 3.

Oliva miniacea (Röding, 1798). Localities: CSI. Figured by Wilson and Gillett (1974), 106: Plate 69, no. 8.

FAMILY MITRIDAE

Mitra cucumerina Lamarck, 1811. Localities: CSI. Figured by Cernohorsky (1976), 296: Plate 256, Figs 13-15; 40: Plate 338, Fig. 1.

Mitra imperialis Röding, 1798. Localities: CH. Figured by Cernohorsky (1976), 293: Plate 255, nos 15, 16.

Mitra tabanula Lamarck, 1811. Localities: CH, CI. Figured by Cernohorsky (1976), 296: Plate 256, no. 22.

Mitra variabilis Reeve, 1844. Localities: PW, CH. Figured by Cernohorsky (1976), 293: Plate 255, Figs 5, 6.

Pterygia barrywilsoni J. Cate, 1968. Localities: PW, MOI. Figured by Wilson and Gillett (1974), 114: Plate 74, Fig. 22.

Vexillum vulpecula jukesii A. Adams, 1853. Localities: MOI, CNI, CH. Figured by Wilson and Gillett (1974), 116: Plate 75, no. 11.

FAMILY VASIDAE

Vasum ceramicum (Linnaeus, 1758). Localities: CSI. Figured by Abbott (1959), 16: Plate 1, Fig. 1.

FAMILY HARPIDAE

Harpa articularis Lamarck, 1816. Localities: CSI. Figured by Rehder (1973), 215: Plate 188, Figs 5-7; 250: Plate 217.

Harpa davidis Röding, 1798. Localities: CSI. Figured by Rehder (1973), 211: Plate 187, Figs 4-6.

FAMILY VOLUTIDAE

Amoria damoni Gray, 1864. Localities: CNI. Figured by Wilson and Gillett (1974), 128: Plate 85, no. 1.

Melo amphora (Lightfoot, 1786). Localities: PW, MOI, CNI, CSI. Figured by Wilson and Gillett (1974), 138: Plate 92, no. 4.

FAMILY CONIDAE

Conus capitaneus Linnaeus, 1758. Localities: CSI. Figured by Wilson and Gillett (1974), 146: Plate 98, no. 10.

Conus coronatus Gmelin, 1791. Localities: CH, CNI, CI, CSI. Figured by Wilson and Gillett (1974), 152: Plate 101, no. 12.

Conus ebraeus Linnaeus, 1758. Localities: PW, CI, CSI. Figured by Wilson and Gillett (1974), 150: Plate 100, Fig. 4.

Conus eburneus Hwass in Bruguière, 1792. Localities: CNI. Figured by Wilson and Gillett (1974), 150: Plate 100, no. 2.

Conus flavidus Lamarck, 1810. Localities: CSI. Figured by Wilson and Gillett (1974), 150: Plate 100, no. 11.

Conus geographus Linnaeus, 1758. Localities: CH, CI, CSI. Figured by Wilson and Gillett (1974), 144: Plate 97, no. 1.

Conus glans Hwass, 1792. Localities: CSI. Figured by Wilson and Gillett (1974), 148: Plate 98, no. 1.

Conus monachus Linnaeus, 1758. Localities: CH, CNI, CI, CSI. Figured by Wilson and Gillett (1974), 152: Plate 101, no. 2.

Conus mustelinus Bruguiére, 1792. Localities: CH, CNI, CI, CSI.

Conus omaria Hwass, 1792. Localities: CSI. Figured by Wilson and Gillett (1974) 146: Plate 98, no. 2.

Conus planorbis Born, 1778. Localities: CSI. Figured by Coleman (1975), 113: Fig. 339.

Conus scabriusculus Dillwyn, 1817. Localities: CI, CSI. Figured by Wilson and Gillett (1974), 150: Plate 100, no. 10.

Conus spectrum Linnaeus, 1758. Localities: MOI, CH, CSI. Figured by Wilson and Gillett (1974), 148: Plate 99, no. 13.

Conus striatus Linnaeus, 1758. Localities: PW, CI, CSI. Figured by Wilson and Gillett (1974), 144: Plate 97, no. 5.

Conus terebra Born, 1780. Localities: PW, CSI. Figured by Hinton (1972), 82: Plate 40, nos 4, 5.

Conus tesselatus Born, 1778. Localities: CSI. Figured by Wilson and Gillett (1974), 150: Plate 100, no. 3.

Conus textile Linnaeus, 1758. Localities: PW, CH, CI. Figured by Wilson and Gillett (1974), 150: Plate 100, no. 3.

Conus vexillum Gmelin, 1791. Localities: CI, CSI. Figured by Wilson and Gillett (1974), 144: Plate 97, no. 10.

Conus victoriae Reeve, 1843. Localities: PW, CH, CNI, CI. Figured by Wilson and Gillett (1974), 144: Plate 97, no. 1.

DISCUSSION

This report lists 146 species of molluscs collected during a brief survey of the Admiralty Gulf, Cape Voltaire and the Institut Islands. Only the amphineurans, meso- and neogastropods are treated here; the remaining groups will be reported in subsequent papers. The study area is on a little-known stretch of the coastline of northern Australia. The ranges of many of the reported species are extended substantially and a number are recorded from Western Australia for the first time.

Most collecting time was devoted to the area around Port Warrender. Because of this the species list from Port Warrender is relatively extensive. Other areas such as Corneille Island and Cassini Island were visited only briefly. If more collecting had been undertaken at those localities more species would undoubtedly have been found. Thus the apparent absence of a given species at a particular locality may only be due to the brief time spent collecting there.

A number of habitats were examined during the survey: mangals, intertidal mudflats, rocky shores, sand beaches and coral reefs. A general pattern of distribution emerged from the distribution of the various species collected. The waters at Port Warrender and Middle Osborn Island have a high silt content and are protected from the open sea. Mangrove and other species able to tolerate the high turbidity of the water were found at these localities. The sites visited on Cape Voltaire and the Institut Islands are exposed to the open sea and have a lower silt content in the water. Exposed rocky shores, sand beaches and coral reefs dominate the habitats of these sites. While some species are found in both, the faunas are fairly distinct.

ACKNOWLEDGEMENTS

The people from whom I received assistance are too numerous to mention, but all are appreciated. Dr I.M. Crawford, P. Randolph, and S. Sledge participated in the August trip. S. Sledge was particularly helpful as a diving partner. Dr B.R. Wilson and C.W. Bryce aided in the field work of the October trip. The Port Warrender location is on a mining lease operated by Alumax Inc., which provided a great deal of assistance through the Mitchell Plateau camp manager, M. Marshal.

Identification assistance was received from Dr B.R. Wilson, G. Hansen and J. Hewitt. Technical assistance came from L. Baxter and C.W. Bryce.

REFERENCES

ABBOTT, R.T. (1959)—The family Vasidae in the Indo-Pacific. *Indo-Pacif. Mollusca* **1:** 15-32.

ABBOTT, R.T. (1960a)—The genus *Strombus* in the Indo-Pacific. *Indo-Pacif. Mollusca* **1:** 33-144.

ABBOTT, R.T. (1960b)—The genus *Lambis* in the Indo-Pacific. *Indo-Pacif. Mollusca* **1:** 147-174.

ABBOTT, R.T. (1968)—The helmet shells of the world. (Cassidae). Part 1. *Indo-Pacif. Mollusca* **2:** 7-202.

CERNOHORSKY, W.O. (1972a)—*Marine shells of the Pacific.* Vol. 2. Sydney: Pacific Publications.

CERNOHORSKY, W.O. (1972b)—Indo-Pacific Nassariidae (Mollusca: Gastropoda). *Rec. Auckland Inst. Mus.* **9:** 125-194.

CERNOHORSKY, W.O. (1974)—The taxonomy of some Indo-Pacific Mollusca with a description of a new species. Part 2. *Rec. Auckland Inst. Mus.* **11:** 121-142.

CERNOHORSKY, W.O. (1976)—The Mitridae of the world. Part 1. The subfamily Mitrinae. *Indo-Pacif. Mollusca* **3:** 273-528.

COLEMAN, N. (1975)—*What shell is that?* Sydney: Paul Hamlyn.

HABE, T. & KOSUGE, S. (1965)—*Shells of the world in color. Vol. 2. The tropical Pacific.* Osaka: Hoikusha.

HINTON, A. (1972)—*Shells of New Guinea and the central Indo-Pacific.* Brisbane: Jacaranda Press.

IREDALE, T. & HULL, A.F.B. (1927)—A monograph of the Australian loricates. Part 9. *Aust. Zool.* **4:** 339-359.

RADWIN, G.E. & D'ATTILIO, A. (1976)—*Murex shells of the world.* Stanford Univ. Press.

REHDER, H.A. (1923)—The family Harpidae of the world. *Indo-Pacif. Mollusca.* **3:** 207-274.

RIPPINGALE, O.H. & McMICHAEL, D.F. (1961)—*Queensland and Great Barrier Reef shells.* Brisbane: Jacaranda Press.

ROSEWATER, J. (1970a)—The family Littorinidae in the Indo-Pacific. Part 1. The subfamily Littorininae. *Indo-Pacif. Mollusca* **2:** 417-506.

ROSEWATER, J. (1970b)—The family Littorinidae in the Indo-Pacific. Part 2. The subfamilies Tectariinae and Echininae. *Indo-Pacif. Mollusca* **2:** 507-534.

WELLS, F.E. (1978)—Zonation of marine gastropods on a rocky intertidal shore in the Admiralty Gulf, with emphasis on the genus *Nerita. Veliger* **20:** 279-287.

WELLS, F.E. & SLACK-SMITH, S. [1980]—Zonation of molluscs in a mangrove swamp in the Kimberley, Western Australia. *Biological Survey of Mitchell Plateau and Admirality Gulf, Kimberley, Western Australia.* Perth: Western Australian Museum

WILSON, B.R. & GILLETT, K. (1974)—*Australian shells.* 2nd. ed. Sydney: A.H. & A.W. Reed.

Received 27 April 1978 Accepted 3 August 1978 Published 28 February 1981

PART 9

ZONATION OF MOLLUSCS IN A MANGROVE SWAMP IN THE KIMBERLEY, WESTERN AUSTRALIA

F.E. WELLS*

and

S.M. SLACK-SMITH*

INTRODUCTION

Little is known of the marine molluscan fauna of the large stretch of coastline of northern Australia between Derby, Western Australia, and Darwin, Northern Territory. The area is isolated and without roads providing access to the coastline. The great distances involved and the expense of mounting expeditions by sea has left this portion of the coastline virtually untouched by marine biologists. The 1976 joint museums' survey provided an opportunity to conduct a field programme in the Admiralty Gulf, about half way between Derby and Darwin. This paper is the first step in documenting the molluscan fauna of that part of the coastline of northern Australia.

During the last few years the Western Australian Museum conducted a number of multi-purpose surveys in the north-eastern corner of the State and the information accumulated allowed Museum biologists to pinpoint areas of particular interest for study. During the course of the October 1976 expedition to the Mitchell Plateau-Admiralty Gulf area the Western Australian Museum Department of Malacology conducted 3 marine programmes. An overall survey of the molluscan fauna of Admiralty Gulf and the nearby Institut Islands was made**. A detailed examination of the patterns of zonation of gastropods on a rocky shore at Walsh Point in the Gulf has been completed (Wells, 1978). The present report deals with the results of an examination of the zonation of molluscs in a mangal at Port Warrender Bay, Admiralty Gulf.

Mangals are one of the dominant marine communities along most of the tropical coastlines of the world, and they are a key feature of the coastline of northern Western Australia. Mangrove tree species are zoned within the swamp according to shore height and distance from the sea or tidal creeks (Macnae 1967, 1968; Wells,

---

*Department of Malacology, Western Australian Museum, Francis Street, Perth, Western Australia. 6000.

**(Part 8, this publication, deals with chitons and meso- and neogastropods.)

Part 3 this report). The invertebrates of the swamps are also zoned (Berry, 1963; Macnae, 1968). Early investigators believed that the molluscs of mangals were peculiarly adapted to the rigors of the community but later studies have shown that most molluscan species are also found on adjacent rocky shores or mudflats and are not restricted to swamps. The most prominent exception is the periwinkle *Littorina scabra* (Linnaeus, 1758) which has a circumtropical distribution but is limited almost exclusively to mangals (Coomans, 1969; Gallagher & Reid, 1974). The molluscan fauna of mangals has been studied in a number of areas geographically close to Western Australia. These include South Africa (Brown, 1971), Queensland (Macnae 1967, 1968), and Malaysia and Singapore (Vermeij, 1973; Berry, 1975). Semeniuk (pers. comm.) is investigating the geological processes occurring in mangals and the zonation of fiddler crabs (*Uca*) at Broome, W.A., but no investigation of the molluscs of mangals has been undertaken in the western half of the Australian continent. This paper on the mangals of Port Warrender, Admiralty Gulf, is intended to help fill that gap.

METHODS

The mangal was divided into 6 zones based on the species of mangrove trees present. Four quadrats 5 m × 5 m were marked off in each zone. The quadrats started at the north end of the mangal and were placed at about 100 m intervals. All species of mangroves in a quadrat were identified according to a key of the Port Warrender mangroves (Semeniuk *et al*, 1978). The trees in a quadrat were counted and the number of each species was recorded. Tree heights were estimated, not directly measured. The molluscs present in each quadrat were divided into 3 groups: those found on the mangroves, individuals found on the mud surface, and those collected in the substrate sample. The substrate sample was obtained by digging an area 50 cm square in the quadrat to a depth of 30 cm and hand sorting the material for molluscs. The mud was too fine to be sieved and logistical problems prevented us from moving it out of the mangal. All molluscs found in each of the 3 areas were collected, tentatively identified, counted, and preserved in 10% formalin buffered with borax. Voucher specimens of all species of molluscs encountered in the study have been deposited in the Western Australian Museum where they have catalogue numbers WAM 482-77 to 500-77.

A line transect was run through the entire mangal to determine the width of each zone. Height of the mangal above datum was determined by measuring water depth every 15 minutes over an entire tidal cycle from low water to low water and comparing the results with the tide charts published for Malcolm Island 5 km away.

RESULTS

The mangrove trees at Port Warrender Bay are located at the south-western corner of Walsh Point (14°34′S, 125°50′E) at the head of the small bay (see also A.G. Wells, Part 3 this publication). They stretch in a band 300 m wide from Walsh Point to the next, unnamed headland 3.2 km south. The mangroves occur at a mid-tidal level of +2.9 to +5.2 m. The maximum tidal range in the area is from −1.7 to +7.1 m. All stations were made at the northern end of the mangal (**Fig. 1**). A tidal flat extends 0.6 km in front of the mangroves at extreme low water, and an extensive mudflat behind the mangroves is 0.8 km wide. The only major tidal channel in the mangal occurs at the southern end and did not influence the zonation patterns in the area sampled.

Mangroves also occurred in a zone 5-10 m wide along the landward margins of the mudflat. Several species of molluscs were found in qualitative examinations of the fringing mangroves: *Cassidula angulifera* (Petit de la Saussaye, 1841), *Cerithidea obtusa* (Lamarck, 1822), *Ellobium aurisjudae* (Linnaeus, 1758), *Littorina scabra*, and *Nerita lineata* (Gmelin, 1791). The two ellobiid species were not found elsewhere in the mangal. Both were buried under a mat of decaying mangrove leaves and branches that did not occur elsewhere in the mangal. *Cassidula angulifera* was found under the mat in open areas between trees while *E. aurisjudae* was half buried in the mud under leaf litter that had accumulated between rocks.

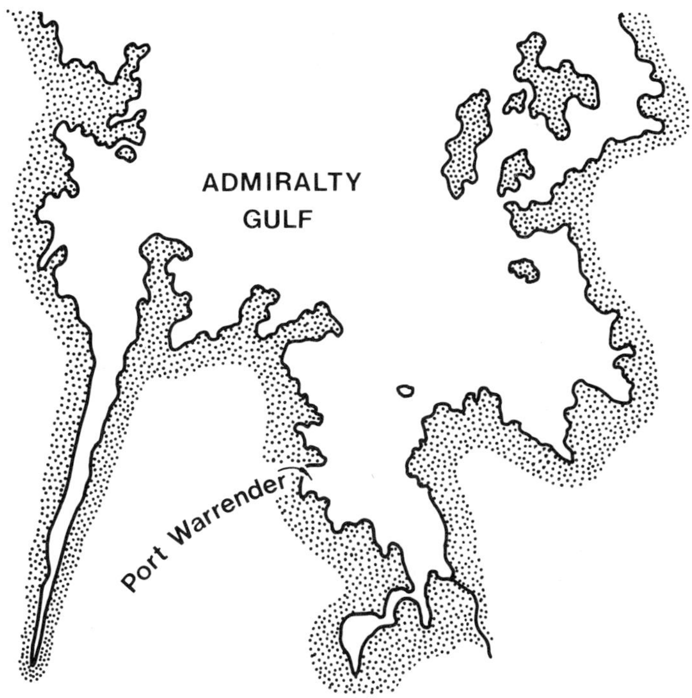

Fig. 1: Map of the Admiralty Gulf showing the position of the mangal at Port Warrender.

The mangal zones

The mangal was divided into 6 zones. Each will be discussed in turn, starting from the landward mudflat and moving seaward:

Zone A: The Mudflat

The extensive mudflat behind the mangroves is unique to arid regions and does not occur in most mangals. The flats are covered only by spring tides and are continuously exposed during the neap tide period. At this time the mud is baked dry by the sun; air temperatures in the area during the study period, 16 October to 6 November 1976, reached as high as 45°C. Although the mud surface was baked dry, salt water seeped readily into holes dug for the substrate samples. Overnight the water depth in holes 30 cm deep reached a depth of 10 cm. No live molluscs were found in the mudflat zone, neither on the mud surface nor burrowed into the mud. The only living animals seen were fiddler crabs and mudskippers.

Zone B: *Ceriops* zone

The 115 m wide *Ceriops* zone was the uppermost mangrove zone. The 4 sampled areas were dominated by a total of 246 trees of *Ceriops tagal* 2-3 m high. Thirteen *Avicennia marina* and 8 *Rhizophora stylosa* 4-5 m high were also encountered in the quadrats. Individuals of *Bruguiera parviflora* were seen in the zone, but none occurred in the quadrats. The trees of this zone were generally short and closely packed together. *Cerithidea obtusa* was the most abundant mollusc, with a density of $1.65/m^2$. They occurred on tree trunks from the mud level to 2 m above the mud surface, and were attached by mucusto the outer lip of the shell aperture in a pattern similar to that described for *Littorina irrorata* (Say, 1822) by Bingham (1972). *Littorina scabra* was found on the trees at a density of $0.13/m^2$. Other species occurred at very low densities: *Nerita lineata* on the trees at $0.02/m^2$, and on the mud surface were an unidentified mytilid sp. ($0.03/m^2$), *Neritina* sp. ($0.02/m^2$), and *Telescopium telescopium* (Linnaeus, 1758) ($0.01/m^2$). No live molluscs were found in the substrate sample.

Zone C: *Aegialitis annulata* zone

Ceriops tagal 2-3 m high were the most numerous mangroves in this zone (440 individuals), but 312 individuals of *Aegialitis annulata* about 1 m high were recorded in the same area. Since no individuals of this species were found in the *Ceriops* zone, its presence clearly demarcated a separate zone. Numerous (78) specimens of *Avicennia marina* 5-7 m high were found in the quadrats of this zone, as were 14 small (1-3 m) *Rhizophora stylosa.* In general the trees were the same size as those in the *Ceriops* zone but were packed more closely together. The zone was 35 m wide.

The molluscan fauna of the *Aegialitis* zone was substantially different from that found in the *Ceriops* zone. *Littorina scabra* was more dense in the trees ($0.35/m^2$), but *Cerithidea obtusa* was substantially less abundant ($0.95/m^2$). *Nerita lineata* were more abundant on the bases of trees, where they had a density of $0.41/m^2$. *Terebralia sulcata* (Born, 1778) were common in the mud of the *Aegialitis* zone but were not encountered in any other zone. Two specimens of the ellobiid *Cassidula* cf *nucleus* (Martyn, 1784) were found in hollow trees. No molluscs were obtained in the substrate sample.

Zone D: *Rhizophora* zone

The most impressive zone in the mangal was the *Rhizophora* zone, in which the only mangrove was *Rhizophora stylosa.* The trees were massive, reaching heights of 10-15 m. The intertwined prop roots themselves reached 1-2 m above the mud surface. Only 4 trees were sampled, 1 to each quadrat.

A few individuals of *Littorina scabra* were found on prop roots and trunks, but the density of this species (0.07/m^2) was less than in the 2 previous zones. It is likely that snails were too high in the trees to be adequately sampled. *Nerita lineata*, which occurred on trees from mud level to 2 m above it, was apparently the dominant mollusc in this zone (0.82/m^2). *Pterynotus permaestus* (Hedley, 1915) (0.13/m^2) occurred along the under sides of the prop roots. Several additional species were encountered at low densities: on the trees were *Isognomon* cf *vitrea* (Reeve, 1858) (0.01/m^2), and *Cerithidea obtusa* (0.01/m^2), and on the substrate were *Onchidium daemelli* (0.02/m^2) and *Telescopium telescopium* (0.01/m^2). There were no live molluscs collected in the sediment sample.

Zone E: *Sonneratia* zone

The mangroves seaward of the *Rhizophora* zone are widely spaced, with large open areas between each tree. This is the most open area of the mangal. Two zones were distinguished seaward of the *Rhizophora* zone, the first of which is the *Sonneratia* zone. Two species of mangroves were encountered in this zone: 7 *Sonneratia alba* 6-8 m high in the 4 quadrats and one *Avicennia marina* of similar height. The zone was 70 m wide.

The molluscan fauna of the *Sonneratia* zone was diverse but only low numbers of each species were found. *Nerita lineata*, the dominant species of the *Rhizophora* zone, here had a density of only 0.04/m^2. Juveniles of this species were found under the bark of living trees and had a calculated overall density of 0.09/m^2. *Littorina scabra* (0.11/m^2) was also found on the trees. A number of species were found on the substrate in low densities: *Salinator burmana* (Blanford, 1867) 0.14/m^2, *Isognomon* cf *vitrea* 0.05/m^2, *Telescopium telescopium* 0.03/m^2, *Enigmonia aenigmatica* (Sowerby, 1825) 0.02/m^2, *Columbella duclosiana* 0.01/m^2, and a cephalaspidean opisthobranch 0.01/m^2. One bivalve of each of the species *Glauconome* cf *cumingi* (Prime, 1862), and *Macoma* sp. were found in the substrate sample. A small species of *Bembicium* was abundant under the bark of *Sonneratia* trees. The small size of this species and its patchy distribution precluded making a quantitative estimate of its density.

Zone F: *Avicennia* zone

The mangrove *Avicennia marina* was the most widely distributed mangrove species, occurring in all mangrove zones except the *Rhizophora* zone. The species was most prominent at the seaward edge of the mangal where 9 individuals 4-8 m high were found in the quadrats. Four trees of *Sonneratia alba* 6-7 m high were also recorded in the quadrats in this zone, which was 25 m wide. Thus *Avicennia* and *Sonneratia* were the only mangroves occurring seaward of the *Rhizophora* zone.

Molluscs were much more numerous in the *Sonneratia* zone than in the *Avicennia* zone. *Columbella duclosiana* increased from 0.01/m^2 to 0.91/m^2 on the bases of the trees and pneumatophores. *Littorina scabra* (1.37/m^2) was widely distributed. The 2 most abundant species on the mud surface were the amphibolid *Salinator burmana* (0.24/m^2) and the ellobiid *Cassidula* cf *nucleus* (0.20/m^2). An unidentified mytilid had a density of 0.35/m^2, but all specimens came from a single clump found in one quadrat. Species present in small numbers were: *Diala* sp. (0.05/m^2), *Nerita lineata*

($0.04/m^2$), and *Telescopium telescopium* ($0.01/m^2$). No live molluscs were found in the substrate sample.

Zonation of molluscs in the mangroves

Small species such as *Diala* and patchy species such as *Brachidontes* were not effectively sampled by our techniques. Data presented for these species should be regarded as an indication of their minimum density and is not sufficiently detailed for an examination of their zonation patterns. The large species were effectively sampled and confidence placed in the measurements of their abundance is greater. Three patterns of zonation were found in the large species of molluscs inhabiting the mangal.

Cerithidea obtusa (**Fig. 2**) was distributed only in the upper 2 mangrove zones, the *Ceriops* and *Aegialitis* zones. The maximum density of this species ($1.65/m^2$) occurred in the uppermost *Ceriops* zone. Density in the lower *Aegialitis* zone tapered to $0.95/m^2$. One individual was found in the *Rhizophora* zone and none in the lower zones. *Terebralia sulcata* (**Fig. 2**) had an even more restricted distribution although similar. The species had a density of $0.33/m^2$ in the *Aegialitis* zone, but no individuals were located outside that zone.

Nerita lineata occurred in all 5 mangrove zones (**Fig. 2**). The species was rare in both the uppermost *Ceriops* zone and the lower *Sonneratia* and *Avicennia* zones. Large numbers of *Nerita* occurred in the *Aegialitis* zone, but maximum density was in the central *Rhizophora* zone. Density decreased both above and below that zone. The graph of *N. lineata* on **Fig. 2** includes only the adult population, which had a mean shell length of 25 mm. A number of small nerites, which may have been juvenile *N. lineata*, of 3-6 mm, were found in the seaward mangrove zones under the bark of *Sonneratia*. It is interesting that no juveniles were found in the *Rhizophora* zones where adult *N. lineata* are most abundant. It seems likely that the *Sonneratia* bark provides suitable protection for newly metamorphosed juveniles but a differential mortality establishes the adult distributional pattern.

Salinator burmana and *Columbella duclosiana* were both found in the seaward mangrove areas but did not occur above the *Sonneratia* zone. *Salinator* was most abundant in the *Sonneratia* zone and declined slightly in the *Avicennia* zone (**Fig. 2**). One individual of *C. duclosiana* was located in the quadrats of the *Sonneratia* zone and none was found in the levels above the *Sonneratia*. A maximum density of $0.91/m^2$ was recorded for *Columbella* in the lowermost *Avicennia* zone.

The total density of molluscs in the 5 mangrove zones showed no clear pattern (**Fig. 3**). The highest density of $3.17/m^2$ occurred in the lowermost *Avicennia* zone. High densities were also encountered in the 2 upper zones, the *Ceriops* and *Aegialitis* zones. Very low densities occurred in the central *Rhizophora* and lower *Sonneratia* zones. Since the mangrove zones are arranged in a line of increasing tidal heights there was no relationship between mollusc density and intertidal shore level. Diversity as measured by Simpson's (1949) index also showed no clear pattern (**Fig. 3**) and did not parallel the variations in density among the various mangrove zones.

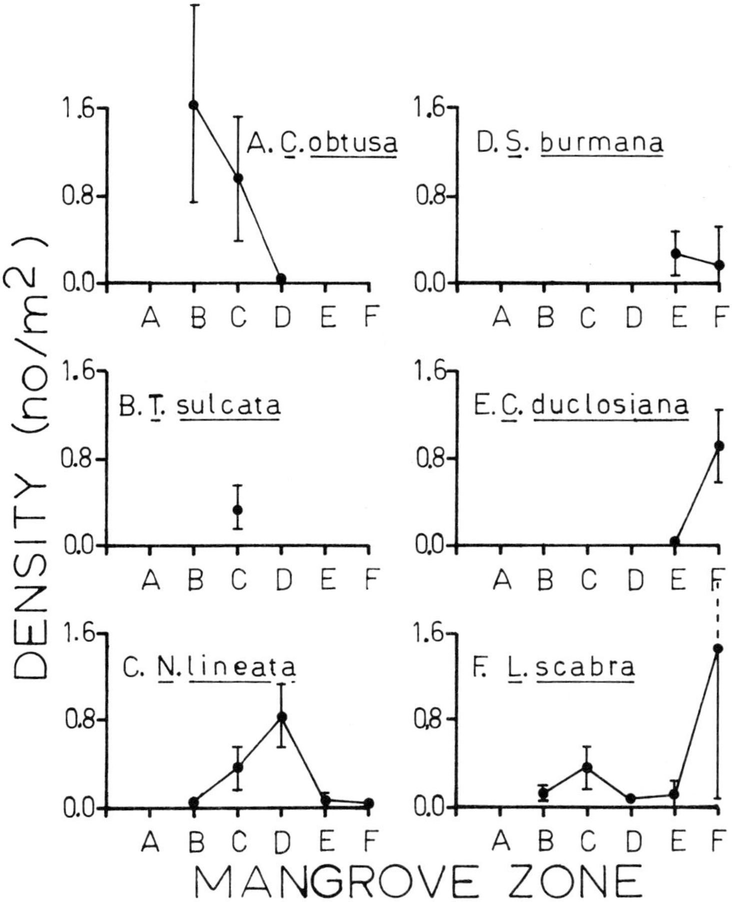

Fig. 2: Densities of selected mollusc species in the various mangrove zones. The mean and one standard deviation from the mean are indicated.

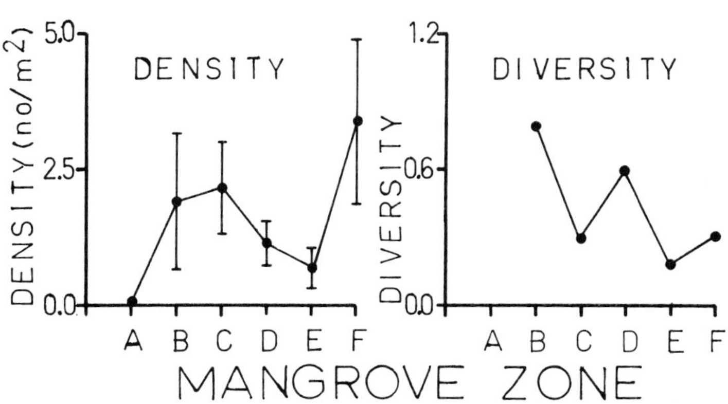

Fig. 3: Density and diversity of molluscs in the various mangrove zones.

DISCUSSION

An increasing number of studies have shown that the Indo-Malayan region has the highest diversity of both mangrove and associated invertebrate species. As the distance from the centre of diversity increases, the number of mangrove and invertebrate species decreases (Macnae, 1968; Brown, 1971). In the centre of diversity Vermeij (1973) recorded 28 species of gastropods from Singapore and Sasekumar (1974) found 25 at Selangor, Malaysia. Sasekumar commented that bivalves were seldom found in the mangals and those that were encountered were not identified. Away from the centre of diversity Macnae (1967) encountered 11 species of molluscs in Queensland and Brown (1971) reported 10 species of gastropods in the Umgeni Swamp at Durban, South Africa. The 14 gastropods and 4 bivalves reported from the mangals of Port Warrender Bay indicate a diversity lower than that of the Indo-Malayan region but higher than the figures reported for South Africa and Queensland.

In general, Caribbean mangroves are less diverse than their Indo-Pacific counterparts (Coomans, 1969). One exception to this statement is the report of 13 species of eolid nudibranchs in a mangal at Port Royal, Jamaica (Edmunds, 1964). No eolids were encountered in the survey at Port Warrender Bay. The Admiralty Gulf mangroves are located on the upper shore levels where they are subjected to prolonged exposure daily. In contrast, the mangroves at Port Royal are in an area where the maximum tidal amplitude is 30 cm and most of the trees are not fully exposed at low tide (Wells, pers. obs.) providing a more suitable habitat for nudibranchs. In addition Edmunds (1964) was collecting in a channel which is continuously covered by seawater. Thus the greater nudibranch fauna of the mangals at Port Royal is a result of hydrographic conditions rather than of inherently greater diversity.

The zonation patterns of both mangrove trees and molluscs in the Admiralty Gulf generally agree with the results of studies conducted in other geographical areas. One difference is the minor importance of the mangrove *Bruguiera* at Port Warrender compared with Queensland. Macnae (1967) described this as a mangrove which predominates in areas with significant freshwater inflows. The diminished numbers of species at Port Warrender can be attributed to the lack of freshwater in the swamp. Berry (1963) distinguished a high tree zone and a low tree zone in mangals at Singapore. At Port Warrender the dominant high tree species was *Littorina scabra*. The dominant species of the lower tree zone varied in the swamp. In the upper zones *Cerithidea obtusa* was most abundant. *Nerita lineata* dominated the central *Rhizophora* zone. The seaward zones were most diverse and no one species was clearly dominant, though *Columbella duclosiana* was the most abundant species. Species living on the mud surface were scattered throughout the mangroves and none was especially abundant.

One of the most fascinating aspects of this study was the complete absence, except for 2 bivalves, of any living molluscs in the substrate itself. This appears to be a general feature of mangals. Berry (1963) found burrowers other than crabs to be virtually absent in the mud. Macnae (1967) reported the bivalve *Geloina coaxans* (Gmelin, 1791), living on the mud surface and suggested that other individuals lived 10-15 cm below the mud surface. Brown (1971) found the gastropod *Auriculastra radiolata* (Morelet, 1860) below the mud surface. Brown commented that he could find no previous literature reference to gastropods living in that habitat. Sasekumar (1974) used a method of sampling the substrate nearly identical to ours and found polychaetes, crustaceans, and sipunculids in profusion. We found individuals of all 3 groups though polychaetes were not abundant. Sasekumar also found no molluscs in the mud. The reasons for the absence of molluscs in the substrate are not clear. Certainly food supplies would be adequate for deposit feeders. The mud is anoxic below a few centimetres but the siphonate bivalves would be able to reach the surface. It is possible that the sediment is too fine for molluscan burrowing though the other 3 animal groups manage. Alternatively the acidity of the substrate could erode the calcareous shells of molluscs and prevent them from surviving in that habitat. Vermeij (1973) pointed out that a large percentage of mangrove molluscs exhibit calcium carbonate resorption or conservation and attributed this to the acidic conditions which prevail in the swamps.

A number of Aboriginal kitchen middens are located on the hills adjacent to the Warrender mangroves. The middens are largely composed of marine molluscan shells and may reach a diameter of 5 m with a height of 1 m. The shells are mainly from the most abundant species of large molluscs in the mangal. An exception to this is the high percentage of valves of *Anadara granosa* (Linnaeus, 1758). This species was not encountered in the quantitative mangrove survey but a few live individuals were collected on the surface of the mudflat seaward of the swamp. The large number of *Anadara* shells in the middens suggests that once the species was more abundant in the area than it is now. The middens have not been dated so we have no idea of the period in which the population declined, but there have been no substantial numbers of Aboriginals in the Admiralty Gulf for the last 40 years (Crawford, pers. comm.). Squires *et al.* (1975) reported that 4 *Anadara* species are sold in the native markets along the Pacific coast of Colombia. The species have an average population density of $1/m^2$ on the mudflats in front of the Colombian mangroves. If the Port Warrender *Anadara* population had a similar density it could have provided the shells found in the middens.

ACKNOWLEDGEMENTS

Participation of the Western Australian Museum in the joint programme with the Field Museum was funded by a special grant from the Treasury Department of Western Australia. The area examined is on a mining lease owned by the Alumax Mining Company. Alumax provided substantial assistance to the programme, especially through the Mitchell Plateau camp manager, M. Marshall. Dr B.R. Wilson headed the expedition and was largely responsible for its success. Dr B.R. Wilson and C.W. Bryce provided field assistance for the mangrove study.

REFERENCES

BERRY, A.J. (1963)—Faunal zonation in mangrove swamps. *Bull. natn. Mus. St. Singapore* **32:** 90-98.

BERRY, A.J. (1975)—Molluscs colonizing mangrove trees with observations on *Enigmonia rosea* (Anomiidae). *Proc. malac. Soc. Lond.* **41:** 589-600.

BINGHAM. F.O. (1972)—The mucus holdfast of *Littorina irrorata* and its relationship to relative humidity. *Veliger* **15:** 48-50.

BROWN, D.S. (1971)—Ecology of Gastropoda in a South African mangrove swamp. *Proc. malac. Soc. Lond.* **39:** 263-279.

COOMANS, H.E. (1969)—Biological aspects of mangrove mollusks in the West Indies. *Malacologia* **9:** 79-84.

EDMUNDS, M. (1964)—Eolid Mollusca from Jamaica, with descriptions of two new genera and three new species. *Bull. mar. Sci. Gulf Caribb.* **14:** 1-32.

GALLAGHER, S.B. & REID, G.K. (1974)—Reproductive behavior and early development in *Littorina scabra angulifera* and *Littorina irrorata* (Gastropoda: Prosobranchia) in the Tampa Bay region of Florida. *Malacol. Rev.* **7:** 105-126.

MACNAE, W. (1967)—Zonation within mangroves associated with estuaries in North Queensland. *Publs Am. Ass. Advmt Sci.* no. 83: 432-438.

MACNAE, W. (1968)—A general account of the fauna and flora of the mangrove and forests in the Indo-West Pacific region. *Adv. mar. Biol.* **6:** 73-270.

SASEKUMAR, A. (1974)—Distribution of macrofauna on a Malayan mangrove shore. *J. anim. Ecol.* **43:** 51-69.

SEMENIUK, V., KENNEALLY, K.F. & WILSON, P.G. (1978)—Mangroves of Western Australia. *Handbk West. Aust. Nat. Club* no. 12.

SIMPSON, E.H. (1949)—Measurement of diversity. *Nature, Lond.* **163:** 688.

SQUIRES, H.E., ESTEVEZ, M., BARONA, O. & MORA, O. (1975)—Mangrove cockles, *Anadara* spp. (Mollusca: Bivalvia) of the Pacific coast of Colombia. *Veliger* **18:** 57-68.

VERMEIJ, G.J. (1973)—Molluscs in mangrove swamps: physiognomy, diversity, and regional differences. *Syst. Zool.* **22:** 609-624.

WELLS, F.E. (1978)—Zonation of marine gastropods on a rocky intertidal shore in the Admiralty Gulf, Western Australia, with emphasis on the genus *Nerita. Veliger* **20:** 279-287.

Received 31 August 1977 Accepted 3 August 1978 Published 28 February 1981